SURFACE CHARACTERISTICS OF FIBERS AND TEXTILES

SURFACTANT SCIENCE SERIES

ADDITIONAL VOLUMES IN PREPARATION

Analysis of Surfactants: Second Edition, Revised and Expanded, *Thomas M. Schmitt*

Fluorinated Surfactants and Repellents: Second Edition, Revised and Expanded, *Erik Kissa*

Physical Chemistry of Polyelectrolytes, *edited by Tsetska Radeva*

Detergency of Specialty Surfactants, *edited by Floyd E. Friedli*

Reactions and Synthesis in Surfactant Systems, *edited by John Texter*

Liquid Interfaces in Chemical, Biological, and Pharmaceutical Applications, *edited by Alexander G. Volkov*

Protein-Based Surfactants: Synthesis, Physicochemical Properties, and Applications, *edited by Ifendu A. Nnanna and Jiding Xia*

SURFACE CHARACTERISTICS OF FIBERS AND TEXTILES

edited by
Christopher M. Pastore
Philadelphia University
Philadelphia, Pennsylvania

Paul Kiekens
University of Gent
Gent, Belgium

CRC Press
Taylor & Francis Group
Boca Raton London New York

CRC Press is an imprint of the
Taylor & Francis Group, an **informa** business

CRC Press
Taylor & Francis Group
6000 Broken Sound Parkway NW, Suite 300
Boca Raton, FL 33487-2742

First issued in paperback 2019

© 2001 by Taylor & Francis Group, LLC
CRC Press is an imprint of Taylor & Francis Group, an Informa business

No claim to original U.S. Government works

ISBN-13: 978-0-8247-0002-7 (hbk)
ISBN-13: 978-0-367-39786-9 (pbk)

Visit the Taylor & Francis Web site at
http://www.taylorandfrancis.com

and the CRC Press Web site at
http://www.crcpress.com

Preface

This book reveals the expanding opportunities for fibers in a wide range of industrial applications. No longer limited to apparel and home furnishings, fibers are being used in medical devices, in aircraft components, and as intelligent sensors. For all these applications, the fiber surface plays an important and fundamental role.

The traditional textile industry needs to understand how fiber surface affects friction, dyeing, wrinkling, and other performance characteristics to optimize production. Newly developing markets such as biomaterials, aerospace, and the automotive industry are interested in more complex performance criteria such as permeability, stiffness, and strength. These properties are also governed to a large extent by the surface of the fiber. This should be no surprise because the high ratio of surface area to volume is a large part of what makes fibers unique.

The topics addressed in this text range from commodity to innovation. The book begins with a discussion of the importance of fiber surface to the traditional textile industry. Following this, novel fibers and their applications are considered. The remainder of the book deals with the ability of fibers to function within composite materials.

The first chapter is, naturally enough, a discussion of cotton fibers—the stalwart of textile fibers. In this chapter, we learn about new techniques for developing wrinkle-resistant finishes on the surface of the fiber using environmentally friendly techniques. This is followed by a discussion of the surface characteristics of polyester fibers—a strong market competitor to cotton. These two chapters alone address the vast majority of textile fiber consumption.

The next two chapters address fundamental issues on the role of fiber surface. The frictional behavior of textiles is described in terms of the fiber surface properties in Chapter 3, and the infrared absorption characteristics (essential for environmental stability, rapid drying, and others) are addressed in Chapter 4.

New fibers and their applications are presented in the next three chapters. A new function for fiber surface is addressed in Chapter 5—the use of fibers as electrochemical sensors in bleaching operations. Chapter 6 discusses the properties of mineral-filled polypropylene fibers. Such fibers are of interest in biomedical applications such as bone plates. High-performance ceramic fibers are described in Chapter 7. These fibers, which are useful for high-temperature applications, typically have very high tensile moduli.

Plasma treatment of fibers is addressed in Chapter 8. Through plasma treatment operations it is possible to dramatically change the surface characteristics of fibers. One typical use is to make the fibers more chemically reactive for subsequent finishing treatments. This includes improving the bonding strength of fibers in resin.

The use of fibers in composite materials is discussed in the final three chapters. Chapter 9 addresses the role of the fiber–resin interface in composites. The interfacial strength of the composite plays a significant role in the strength and damage tolerance of these advanced materials.

Chapter 10 addresses the role of fiber surface on the thermal properties of composite materials. Chapter 11 presents a new concept in permeable composites. These materials may have traditionally been seen as inadequate for structural applications, but have function in interesting areas such as acoustic baffles in aircraft engines.

It is exciting to find so many different and exciting opportunities for fibers! In all the applications presented in this text, fiber surface plays an important role.

Christopher M. Pastore
Paul Kiekens

Contents

v

Contributors

Marie-Hélène Berger   Centre des Matériaux, Ecole des Mines de Paris, Evry, France

Anthony R. Bunsell   Centre des Matériaux, Ecole des Mines de Paris, Evry, France

Wallace W. Carr   School of Textile and Fiber Engineering, Georgia Institute of Technology, Atlanta, Georgia

Matthew Dunn   Fiber Architects, Maple Glen, Pennsylvania

Brian George   School of Textiles and Materials Technology, Philadelphia University, Philadelphia, Pennsylvania

Yasser A. Gowayed   Department of Textile Engineering, Auburn University, Auburn, Alabama

Bhupender S. Gupta   Department of Textile Engineering, Chemistry, and Science, North Carolina State University, Raleigh, North Carolina

You-Lo Hsieh   Department of Textiles, University of California at Davis, Davis, California

Samuel Hudson   Department of Textile Engineering, Chemistry, and Science, College of Textiles, North Carolina State University, Raleigh, North Carolina

Paul Kiekens   Department of Textiles, University of Gent, Gent, Belgium

Marian G. McCord   Department of Textile Engineering, Chemistry, and Science, College of Textiles, North Carolina State University, Raleigh, North Carolina

Elizabeth G. McFarland   School of Textile and Fiber Engineering, Georgia Institute of Technology, Atlanta, Georgia

Cezar-Doru Radu   Department of Textile Finishing, Technical University of Iasi, Iasi, Romania

D. S. Sarma   Trident, Inc., Brookfield, Connecticut

Peter Schwartz   Department of Textiles and Apparel, Cornell University, Ithaca, New York

Eduard Temmerman   Department of Analytical Chemistry, University of Gent, Gent, Belgium

Jo Verschuren   Department of Textiles, University of Gent, Gent, Belgium

Clark M. Welch   Southern Regional Research Center, Agriculture Research Service, U.S. Department of Agriculture, New Orleans, Louisiana

Philippe Westbroek   Department of Analytical Chemistry, University of Gent, Gent, Belgium

1

Formaldehyde-Free Durable Press Finishing

CLARK M. WELCH Southern Regional Research Center, Agriculture Research Service, U.S. Department of Agriculture, New Orleans, Louisiana

I. INTRODUCTION

Chemical treatments applied to cellulosic textiles to impart wrinkle resistance, permanent creases, shrinkage resistance, and smooth drying properties are often referred to as durable press (DP), easy care, or wash–wear finishes. Such dimensional stabilization or shape fixation processes are applied to yarns, fabrics, or entire garments made of cotton or its blends with polyester. DP finishes are also applied to textiles of rayon or other forms of regenerated wood cellulose. Occasionally, fabrics of linen and ramie are treated. Chemical finishes for wool, a protein fiber, have made launderable wool garments a commercial reality. The DP finishing of silk, likewise a protein material, has been the subject of promising recent research and development [1].

The method used in DP finishing of cellulosic textiles is to apply a cross-linking agent that reacts with hydroxyl groups of cellulose in the presence of heat and catalysts to form covalent cross-links between adjacent cellulose molecular chains. Because cotton cellulose is a high polymer with molecular weights exceeding 1 million, multiple cross-linking creates a three-dimensional network within each fiber. The fibers, yarns, and fabrics so treated exhibit increased resilience. When bent, flexed, or otherwise deformed during garment use or laundering, the fabric returns to the flat or creased configuration it possessed at the time the cross-links were put in place.

Formaldehyde is easily the least expensive, most effective cross-linking agent known for cellulose and proteins, but is also an irritant and a mutagen in certain bacterial [2,3] and animal species [4–6], and it is officially classified as a probable human carcinogen [6]. Limits have been established in many countries with regard to the maximum concentrations of formaldehyde vapor that are allowable in the workplace over various exposure times [7,8]. Formaldehyde readily adds to amides and ureas to yield N-methylol agents that are highly effective cross-linking agents and that cause less fabric strength loss during treatment than does formaldehyde itself. However, the fabrics treated with such agents gradually and continually release free formaldehyde over indefinite periods. Alkyl or hydroxyalkyl ethers of N-methylol agents are the DP finishing agents currently in widespread use, because they have much lower formaldehyde release [9,10], but such "capped" agents also have decreased effectiveness as finishing agents. Because of the formaldehyde-release problem, periodic medical testing of exposed workers is required. To the cost of medical examinations and medical recordkeeping is added the cost of air monitoring and ventilation to keep formaldehyde levels at or below the maximum set by law. The buildup of formaldehyde vapor in the unused space of closed containers of finished goods is a continuing problem during shipping or storage.

The mutagenic and cytotoxic activity of a variety of aldehydes, including formaldehyde, has been noted and mechanisms proposed which involve lipid peroxidation [11]. The possibility exists that aldehydes as a class may present some health risks as textile finishing agents.

The need for formaldehyde-free DP finishing agents is twofold:

1. To decrease or eliminate possible health risks associated with the use of agents that contain formaldehyde or release it during the steps of fabric finishing in the textile mill, fabric cutting and sewing into garments, shipping and storage of finished goods, textile retailing, and for two or three launderings [12] in final use of the fabrics or garments by the consumer
2. To develop alternative finishing agents that may provide new or improved textile properties less readily obtainable in conventional finishing processes.

Primary emphasis in this survey is placed on formaldehyde-free agents that react in the interior of cotton fibers to produce cross-linking and dimensional stabilization. However, fiber surface treatments with polymeric agents that are self-cross-linking and elastomeric, are grafted to the surface of the fibers, or those which serve as fiber lubricants to enhance fabric softness and recovery from wrinkling, are important aids in DP finishing. The concurrent use of such auxiliary agents often decreases the amount of cellulose cross-linking needed, thus reducing the tensile and tearing strength losses caused by cross-linking. The presence of excessive cross-linking inherently prevents even distribution of applied stress between the load-bearing elements within fibrils and microfibrils in cotton fibers, thus leading to fiber embrittlement and fabric tendering.

To be suitable for DP finishing, a formaldehyde-free cross-linking agent must be stable in water solution and soluble to the extent of 15–20% by weight so as to make energy-saving low, wet pickup treatments possible. The agent should remain colorless and nonvolatile during heat curing at 150–200°C in the presence of environmentally acceptable catalysts that do not degrade cellulose at the curing temperature used. To be suitable for high-speed, continuous fabric processing, the agent should give the necessary level of cellulose cross-linking in oven residence times of 10–20 s. This requirement does not necessarily apply to garment curing, which is a batch operation and conventionally may take 10–20 min at 145–155°C, as, for example, in treating trousers [13]. The candidate agent and any vapors from the resulting finish should be less irritating, odorous, toxic, or mutagenic than is the case for conventional formaldehyde-derived agents or finishes. The finish should continue to impart the needed level of DP performance through 20–50 home launderings, depending on the type of garment, bedding, or household fabrics involved. Finally, the candidate agent and catalyst should be widely available at low cost.

Although no single DP finishing agent is known which meets all of these requirements at the present time, a variety of formaldehyde-free cellulose cross-linking agent and combinations of these have been studied. Low-molecular-weight diepoxides and triepoxides, as well as divinyl sulfone and its adducts, currently appear to be excluded because of mutagenicity, toxicity, or lachrymatory properties.

II. GLYOXAL

A. Early Studies

This dialdehyde, also named ethanedial, is important as a low-cost highly water-soluble, highly reactive raw material for the manufacture of 1,3-dimethylol-4,5-dihydroxyethyleneurea (DMDHEU) and its alkyl or hydroxyalkyl ethers. The latter ''capped'' agents are the principle DP finishing agents in current use. Glyoxal itself reacts with cotton in the presence of acid catalysts to produce cellulose cross-linking [14]. The reaction has been depicted as follows:

$$O{=}HC{-}CH{=}O + 2\ cell{-}OH \rightarrow O{=}HC{-}CH(O{-}cell)_2 + H_2O$$

$$O{=}HC{-}CH(O{-}cell)_2 + 2\ cell{-}OH$$

$$\rightarrow (cell{-}O)_2HC{-}CH(O{-}cell)_2 + H_2O$$

In these equations, ''cell'' is a portion of a cellulose molecular chain. The first aldehyde group of the glyoxal molecule often couples with two hydroxyl groups of one cellulose molecule, and the second aldehyde group may couple with two hydroxyl groups of a second cellulose molecule, thus completing the cross-link. A more complicated reaction can involve coupling of the first aldehyde group with two different cellulose molecules. The second aldehyde group may then couple with the same two cellulose molecules, or with a third one.

Magnesium chloride has been used as a catalyst in pad-dry-cure treatments of cotton fabrics with 10% solutions of glyoxal in the presence and absence of active hydrogen compounds as additives. The latter were covalently bonded to the cellulose by the glyoxal [15]. A medium level of wrinkle resistance was imparted. The additives greatly increased the fabric weight gains, but led to greater strength losses. Increased moisture regain relative to fabric cross-linked without additives was produced by grafting ethylene glycol, glycerol, sorbose, starch, butyramide, polyacrylamide, and tris(hydroxymethyl) phosphine oxide to the cotton. Moreover, the moisture regain equaled or exceeded that of untreated cotton. This is remarkable because conventional DP finishing almost invariably decreases the moisture regain of cotton. Other research [16,17] has shown that the use of magnesium chloride catalysis tends to produce fabric yellowing by glyoxal, as well as severe strength loss. However, a portion of the tendering may result from very short acetal cross-links of the type $[-O-CH(R)O-]$ which glyoxal can form in cellulose. The effective length of such linkages is the same as for monomeric $(-O-CH_2O-)$ linkages produced to some extent by formaldehyde [18].

With aluminum sulfate as the catalyst, glyoxal was observed to impart high levels of wrinkle resistance [17]. Very high conditioned wrinkle recovery angles of 280°–300° (warp + fill) were imparted at glyoxal concentrations of 4.8–15%. Fabric strength retention was extremely low at the lowest glyoxal concentrations, but increased twofold (to 45–56%) at the highest concentrations. Glyoxal present in excess can apparently act as a chelating agent and diluent for aluminum sulfate, thus decreasing the tendency of this catalyst to degrade cotton cellulose during curing at 135–160°C. When polyhydric alcohols were present as additives, increased DP appearance ratings and improved fabric whiteness resulted [17]. Fabric weight gains corresponded to grafting of two-thirds of the applied ethylene glycol to the cotton. Strength losses were, however, prohibitive (60–80%).

B. Glyoxal–Glycol Mild Cure Processes

Considering that cellulose cross-linking, acid-induced cellulose chain cleavage, and oxidative yellowing are competing processes during heat curing of cotton with glyoxal, it appeared desirable to speed up cross-linking as much as possible under mild conditions. By adding an α-hydroxy acid to activate the aluminum sulfate catalyst, and lowering the cure temperature to 115–125°C, considerable strength improvement was obtained. Tartaric and citric acids were the most effective activators. An emulsified silanol-terminated silicone showed synergism with glyoxal and the activated catalyst in imparting increased DP performance. A series of glycols of various chain lengths were compared as additives. The order of overall effectiveness in increasing the DP appearance ratings was ethylene glycol <1,3-propanediol = diethylene glycol < 1,6 hexanediol ≫ triethylene glycol [19–21].

With 1,6-hexanediol as the additive, a nonionic polyethylene was suitable as fabric softener. This avoided the water-repellent effect that a silicone would produce. A formulation containing 2.4% glyoxal, 4.9% 1,6-hexandiol, 0.77% aluminum sulfate 16-hydrate, 0.37% tartaric acid, and 0.25–0.50% polyethylene was developed [21]. Drying the impregnated fabric at 85°C and curing at 120°C for 2 min imparted the following properties to 80 × 80 cotton printcloth : DP rating 4.0–4.2, conditioned wrinkle recovery angle 282°–291° (warp + fill), wet wrinkle recovery angle 238°–246° (warp + fill), tearing strength retention 54–55%, breaking strength retention 51–54%, and bending moment 65% of that for untreated fabric. Thus, the glyoxal–glycol mild cure process imparted properties fairly similar to those obtained in conventional finishing with N-methylol agents.

The role of the added glycol appeared to be that of a cross-link modifier which reacted with hemiacetal formed from glyoxal with cellulose, and thereby increased the spacing, branching, and flexibility of the final three-dimensional cross-link network produced in the cotton fibers.

C. High-Temperature Processes

A mixed catalyst containing aluminum sulfate and magnesium sulfate has been used as a nonyellowing curing agent for glyoxal finishing at 190–205°C in the presence of a reactive silicone, which exerted a synergistic effect [22]. At a level of DP performance equal to that imparted by conventional treatment with DMDHEU, the glyoxal-finished fabric had higher tearing strength and slightly lower breaking strength than the conventionally treated fabric. These treatments were run on 65/35 polyester/cotton twill. Addition of an alkaline buffer such as sodium metaborate to the formulation eliminated the need for an afterwash to remove acidic catalysts [23]. It is customary to use a high curing temperature for the finishing of cotton–polyester blend fabrics, so as to give the maximum rate of fabric throughput and to ensure heat-setting of the polyester component to final fabric dimensions.

The glyoxal–glycol process previously described has been modified for use on all-cotton fabric at medium temperatures (145–160°C) and also for curing at 170°C for 35 s [21]. In the latter instance, the catalyst was 0.5% aluminum chlorohydroxide, $Al_2(OH)_5Cl \cdot 2H_2O$, plus 3% lactic or glycolic acid as an activator. The cross-link modifier was 4% 1,6-hexanediol, used with 0.5% nonionic polyethylene as fabric softener. With 2.4% glyoxal as cross-linker, the DP ratings were 4.0–4.2, conditioned wrinkle recovery angles were 278°–287° (warp +

fill), and tearing strength retention was 62–64%. Breaking strength retention was 44–45%. The whiteness index measured spectrophotometrically was 97% of that for untreated fabric. Judging from the breaking strength retention, this high temperature would be more suitable for fabric containing 65% cotton and 35% polyester.

D. Appraisal of Glyoxal as a Finishing Agent

Glyoxal is mutagenic in a wide range of bacteria, and oral studies indicate it can act as a tumor promoter, but not as an initiator. Clinical studies have given no evidence of sensitization effects. Glyoxal is somewhat irritating to the mucous membrane. Additional data are necessary before the agent can be considered safe for use in cosmetic products [24]. This may also apply to its use in textile finishing.

In addition to problems encountered with decreased breaking strength in glyoxal-treated fabrics, industrial research workers have observed intermittent problems with fabric yellowing by this agent. Cotton grown in soil rich in iron oxide is likely to have more than the usual traces of this compound present in the fibers and appears especially prone to yellowing when finished with glyoxal. Gas-fired curing ovens appear more likely to induce yellowing of glyoxal finishes than electrically heated ovens. In some instances, sunlight can induce discoloration of unwashed samples. These observations suggest the need for an additive that can remove or derivatize any free aldehyde or hemiacetal groups remaining in glyoxal-finished fabrics.

What previous studies have shown is that a high level of wrinkle resistance and smooth-drying properties can be imparted by glyoxal in the presence of glycols. Extended laundering durability studies would be desirable on these finishes. All commercial grades of glyoxal contain some formaldehyde, and it is not certain when formaldehyde-free glyoxal will be available in bulk quantities.

III. OTHER ALDEHYDES

A. Performance Relative to Conventional Agents

A comparison has been made of 11 aldehydes, including formaldehyde, glyoxal, and glutaraldehyde, as DP finishing agents in pad-dry-cure treatments with 1.8–2.0% magnesium chloride hexahydrate as the catalyst [25,26]. Curing was at 160°C for 3 min. Surprisingly, glutaraldehyde was comparable to formaldehyde and DMDHEU in effectiveness, in terms of the conditioned wrinkle recovery angle (274°–284°) imparted. Glutaraldehyde was also comparable to formaldehyde in DP appearance rating imparted and was superior to glyoxal. No fabric softener or other additive was used in these treatments. Glyoxylic acid was fairly effective if applied from water solutions at pH 2, at which the carboxyl group is nonionized, but was not effective at pH 4–5, at which the acid is converted to its salt [25].

The aldehydes that most readily form hydrates in water, as indicated by nuclear magnetic resonance (NMR) spectra, were the most active cellulose cross-linking agents. The first step in the cellulose cross-linking process is in all probability the formation of a cellulose hemiacetal. An aldehyde that readily forms a hydrate is likely to form a hemiacetal also, as both reactions involve addition of hydroxyl group to the carbonyl group of the aldehyde.

Glutaraldehyde exists in water solution primarily as 2,6-dihydroxytetrahydropyran formed by cyclization of glutaraldehyde hydrates [27]. Consequently, this agent probably reacts with cellulose as a difunctional cross-linking agent, rather than as a tetrafunctional agent. Resistance of the glutaraldehyde DP finish to hydrolysis by $0.5N$ hydrochloric acid was less than for formaldehyde, glyoxal, or DMDHEU [26]. For practical uses, the durability of glutaraldehyde finishes to alkaline hydrolysis would be much more important than their stability to acid hydrolysis, as home laundering is generally carried out with alkaline detergent.

In a more recent study [28], 6–8% aqueous solutions of glutaraldehyde were applied to cotton fabric by pad-dry-cure treatment using aluminum sulfate as the catalyst. After a mild cure (135°C for 3 min), extremely high wrinkle recovery angles (295°–304°, warp+fill) were observed. Tensile strength retention was 31–46%. The same treatment using formaldehyde imparted appreciably lower wrinkle recovery and lower strength retention.

B. Appraisal of Alternative Aldehydes as Finishing Agents

Glutaraldehyde and glyoxylic acid are like glyoxal in being mutagenic toward some species of bacteria [29–31]. They are also rather strong irritants and have a considerable odor. Glutaraldehyde appears to be a superior cellulose cross-linking agent, but is several times as expensive as conventional agents. Animal toxicity studies [32,33] suggest care is needed in the handling of glutaraldehyde. Studies on human response to exposure are rather incomplete, however [34–36].

IV. ACETALS OF MONOALDEHYDES AND DIALDEHYDES

A. Types of Acetals Effective

The cross-linking of cellulose by an acetal may be represented as follows:

$$RCH(OR')_2 + 2 \text{ cell}—OH \rightarrow \text{cell}—O—CHR—O—\text{cell} + 2R'OH$$

Magnesium chloride has been used as the catalyst alone or with citric acid as an activator in treatments comparing 16 monoacetals and diacetals in the DP finishing of cotton printcloth [37]. The most effective finishing agents were 1,1,4,4-tetramethoxybutane and 2,5-dimethoxytetrahydrofuran applied in the presence of an activated catalyst. As a rule, the diacetals were noticeably less effective than the dialdehydes from which they were derived. Moreover, the resulting DP finishes were less resistant to acid hydrolysis than the dialdehyde finishes or DMDHEU, although superior in this respect to other types of nitrogenous DP finishes. The diacetals were more effective than monoacetals.

Several dialdehydes have been applied to cotton fabric as preformed reaction products with linear or branched polyols [38] or as dialdehyde–polyol mixtures, along with various polymers of low glass transition temperature (T_g). Among the metal salt catalysts used were halides of magnesium and aluminum. A glyoxal reaction product with pentaerythritol was applied with a butyl acrylate–vinyl acetate copolymer having a T_g equal to $-28°C$, to impart a DP appearance as high as 4.5.

Dimethyl [39] and diethyl [40] acetals of glyceraldehyde have been applied to impart wrinkle resistance to cotton, using aluminum sulfate as a catalyst alone or activated by tartaric acid. Thus, the monomeric carbohydrate, glyceraldehyde, was applied in the form of its acetals to cross-link the polymeric carbohydrate, cellulose. Other α-hydroxyl acetals studied were 2,3-dihydroxy-1,1,4,4-tetramethoxybutane and 3,4-dihydroxy-2,3-dimethoxytetrahydrofuran [39]. The presence of α-hydroxy groups in these four agents should enable them to polymerize concurrently with cross-linking of the cellulose. Fabric yellowing is a problem with this type of agent.

B. Appraisal of Acetal Finishes

When simple acetals are used to cross-link cellulose, alcohols are a coproduct of the reaction, as seen from the chemical equation in Section IV.A. Thus, the DP finishing process can lead to the production of volatile organic compounds (VOCs) that will have to be vented into the atmosphere or condensed in a recovery system connected to the curing oven exhaust. Typically, the alcohol evolved would be methanol or ethanol. On the other hand, if polyols of low volatility are prereacted with glyoxal to form complex acetals, these acetals can be used to produce some cross-linking of cellulose as well as forming polymeric acetals grafted to cellulose. Such finishing processes should not produce appreciable VOCs. The acetals may be regarded as "capped" aldehydes, and the amounts of free aldehydes formed during heat curing should be small. As non-nitrogenous compounds, acetals should yield finishes that are chlorine resistant, but the relatively sluggish reactivity, increased cost, and limited water solubility of many of these agents are disadvantages. A monoacetal of glyoxal having greater interest is 2,2-dimethoxyacetaldehyde [41]. It forms cellulose cross-linking agents by reaction with substituted ureas, as discussed in Section V.G.

V. ADDUCTS OF GLYOXAL WITH UREAS AND AMIDES

A. The Cyclic Adduct with *N,N'*-Dimethylurea

The monomeric addition product of glyoxal with *N,N'*-dimethylurea is 1,3-dimethyl-4,5-dihydroxy-2-imidazolidinone, often called DHDMI. It has also been named 1,3-dimethyl-4,5-dihydroxyethyleneurea and is commercially the most important nonformaldehyde cross-linking agent of the glyoxal–urea type. The first article on DHDMI appeared in 1961 [42] and reported many of the main features of this compound as a DP finishing agent.

In the synthesis of DHDMI, two products were isolated; they were later identified by infrared and NMR spectra as cis and trans isomers [43]. Either an isomer or a mixture of the two could be used in treating cotton fabric. The agent at a concentration of 10% in water was applied with 4% magnesium chloride hexahydrate as a catalyst. Pad-dry-cure treatment imparted a moderate conditioned wrinkle recovery angle (246°, warp + fill) to cotton fabric. The cure was at 160°C for 3 min. No fabric softener was used. After 10 launderings, the wrinkle recovery angle fell to 216°.

Zinc fluoroborate was the most effective catalyst in terms of wrinkle recovery angle (270°–271°) imparted. DHDMI was considerably less effective than dimethylolethyleneurea (DMEU). The cyclic adduct of glyoxal with urea, known as 4,5-dihydroxy-2-imidazolidinone (DHI) or as 4,5-dihydroxyethyleneurea, imparted higher wrinkle recovery angles than DHDMI. However, DHI finishes yellowed during the cure and afterwash, and were susceptible to chlorine damage.

The DHDMI finishes were very highly resistant to chloride damage even after repeated laundering and surpassed DMEU in this property. The DHDMI finishes also were more stable to acid hydrolysis than DMEU treatments, but less resistant than finishes from *N,N'*-dimethylolurea.

The first patent on DHDMI as a finishing agent appeared in 1963 and it likewise disclosed zinc fluoroborate as a particularly effective catalyst [44]. However, this catalyst presents environmental and disposal problems and has not been recommended for commercial use.

The kinetics of DHDMI reaction with cotton at low curing temperatures established that the rate is first order with respect to DHDMI when zinc salts are present as the catalyst. Zero-order reaction occurs with magnesium salts as the catalyst [45]. The presence of *N*-methyl groups, as in DHDMI, increased the rate of reaction with cotton when zinc salts were the catalysts. An S_n1 carbocationic mechanism was indicated for DHDMI and DHI. An S_n2 mechanism was proposed for DMEU reaction with cotton, however [46].

Use of technical grades of DHDMI have led to fabric yellowing during the cure. The presence of certain buffers during the reaction of glyoxal with dimethylurea in preparing DHDMI and subsequent addition of an alcohol and acid to alkylate the hydroxyl group of DHDMI has been recommended as a way of avoiding the yellowing problem without having to isolate and purify the DHDMI prior to use [47]. The DP performance imparted by diluted crude product, after the addition of a catalyst, was also improved over the results with crude DHDMI as ordinarily prepared.

The addition of 5% acrylate copolymer having a T_g of $-20°C$ to 10% DHDMI and 1% zinc fluoborate catalyst has been found to increase the wrinkle recovery angle and DP appearance rating imparted to cotton printcloth. Performance levels comparable to DMDHEU finishes were observed [48]. The tearing strength retention was not improved by the acrylate copolymer, but was improved when polyethylene and silicone were also added. A progressive increase in conditioned and wet wrinkle recovery angle occurred as the T_g of the added copolymer was decreased [49]. Treating formulations gave negative Ames tests for mutagenicity with Salmonella TA bacteria. Treated fabrics gave nearly negative skin patch tests.

A characteristic of certain formaldehyde-free finishing agents is that they give a less uniform distribution of cross-links in the cotton fiber than do conventional *N*-methylol agents [50,51]. DHDMI apparently cross-links cellulose chains within the lamellae, but does not form cross-links between lamellae. This was demonstrated by the extensive layer separation which occurred in DHDMI-treated fibers subsequently embedded in a methacrylate polymer and swollen, as observed by electron microscopy on fiber cross sections. Because the DHDMI-treated fibers were almost completely insoluble in cupriethylenediamine hydroxide, cellulose cross-linking had occurred and was evidently within the lamellae.

The moisture regain and the affinity for CI Direct Red 81 of cotton cross-linked by DHDMI were the same as the untreated cotton [50]. The water of imbibition was decreased by DHDMI finishing, although not as much as by DMDHEU. Recovery from tensile strain was the same for the two finishes. Fabric breaking strength retention at a given level of wrinkle resistance was greater for DHDMI than DMDHEU.

The DP finishing of cotton with conventional *N*-methylol agents greatly decreases the receptivity and affinity of the cotton for direct dyes. This so-called dye resist is much less evident in fabric finished with DHDMI [52]. When the molecular weight of the dyes is as low as 600, approximately, such dyes have about the same affinity for DHDMI-treated fabric as for unfinished cotton [53]. The afterdyeing of wash−wear garments is made possible in this case [52,54]. It was found necessary to dye the material at pH 8.0 rather than in solutions acidified to pH 3.0 with aqueous acetic acid, as even this mild acidity stripped most of the DHDMI finish [53]. Magnesium chloride was the DP finishing catalyst used.

By means of reverse gel permeation chromatography, it has been found that in cotton fabric cross-linked with DHDMI, the fibers retain greater accessible internal volume over the entire range of pore size than do fibers cross-linked with DMDHEU [55]. Increasing the add-on of DHDMI raised the accessible internal volume, whereas increasing the add-on of DMDHEU caused a further decrease in this property.

Postmercerization of DHDMI-finished cotton fabric has been shown to enhance subsequent afterdyeing [53,54]. Direct dyes of molecular weight too high for use without this postmercerization step were strongly absorbed and gave deeper dyeing than on unfinished cotton or even on mercerized cotton.

The amine odor and off-white color sometimes evident in DHDMI-finished fabric can be prevented in some cases by the use of coreactive additives such as polyhydric alcohols in the treating formulation [56,57]. Apparently, these additives alkylate hydroxyl groups of DHDMI. Preformed ethers of DHDMI with simple alcohols [58] and polyols [59] have been recommended as finishing agents.

B. Appraisal of DHDMI as a DP Finishing Agent

Although DHDMI has been used to a moderate degree as a commercial nonmutagenic DP finishing agent, a recent attempt to make it a major factor in DP finishing was unsuccessful because of technical difficulties. Odor and yellowing problems and the need for environmentally harmful catalysts such as zinc fluoroborate in order to achieve a satisfactory level of DP performance have been continuing obstacles. Currently recommended catalysts are proprietary in nature, and little recent work has been published on the performance of catalysts of specified composition. The use of purified DHDMI results in satisfactory whiteness in treated fabric, but the cost of the purified agent is about double the cost of conventional finishing agents, taking into account the high DHDMI concentrations required in the finishing formulations.

C. Polymeric Adducts of Glyoxal with Urea

Oligomers of glyoxal with cyclic ethyleneurea are formed when the two compounds are reacted in water solution at a 1:1 mole ratio at pH 8 [60]. The average molecular weight of the oligomers was slightly more than double the sum of the molecular weights of glyoxal and ethyleneurea. From NMR spectra and chromatographic data, it was concluded that the oligomers contain alternate glyoxal and ethyleneurea units. DP finishing with the crude reaction mixture caused little fabric discoloration. With magnesium chloride as the catalyst, the finishing process imparted higher conditioned wrinkle recovery angles but lower DP appearance ratings than DHDMI. The oligomer finish was more resistant to acid hydrolysis than that from DHDMI.

Later studies indicate the reaction of glyoxal and ethyleneurea is almost quantitative. The main products are linear oligomers with an average molecular weight of about 600 [61]. The copolymer can be further reacted with methanol to produce a methylated oligomer imparting a degree of wrinkle recovery together with improved strength retention [62]. Formation of a glyoxal–ethyleneurea oligomer in the presence of diethylene glycol or other polyols has also been carried out to form a suitably modified finishing agent [63].

Glyoxal has been observed to react with 1,1'-ethylenebis(3-methylurea) in aqueous solution to yield a tetrafunctional cross-linking agent that could be isolated with some difficulty. The product was shown to be 1,3-bis(4,5-dihydroxy-3-methyl-2-oxoimidazolidin-1-yl)ethane [64]. Its effectiveness as a cross-linking agent was almost identical with that of DHDMI. The results show that the beneficial effect of doubling the number of cellulose-reactive groups was just offset by the adverse effect of doubling the size of the molecule. A cross-linking agent cannot be very effective if its molecules are too large to penetrate well into the cotton fiber interior.

D. Assessment of Polymeric Glyoxal–Urea Adducts as Finishing Agents

To avoid fabric yellowing, it appears necessary that the adduct be formed from an N,N'-disubstituted urea, which may be linear or cyclic. Ethyleneurea is far more expensive than urea, and the polymeric adducts studied so far do not have offsetting advantages in performance as DP finishing agents. Polymeric adducts of glutaraldehyde with urea have proven to be less effective agents than DHDMI [65].

E. Adducts of Dialdehydes with Amides

Based on NMR spectra, the reaction product of equimolar amounts of glyoxal and methyl carbamate, as formed in water solution, was bis(carbomethoxy)-2,3,5,6-tetrahydroxypiperazine [66]. This reagent imparted higher wrinkle recovery angles but lower DP appearance ratings than DHDMI. In a similar manner, the 1:1 adduct of glyoxal with acrylamide was formed in water solution and characterized by NMR as 1,4-diacryloyl-2,3,5,6-tetrahydroxypiperazine [67]. At very high concentrations (15–17%), it imparted considerably higher wrinkle recovery angles than DHDMI and the same or slightly higher DP performance. The finish had greater resistance to acid hydrolysis than DHDMI, but lacked chlorine resistance, and produced fabric yellowing during the cure.

Glutaraldehyde and methyl carbamate reversibly form a 1:1 adduct in aqueous solution. As a DP finishing agent, this adduct produced some fabric discoloration, although less than the 1:1 adduct from glutaraldehyde and urea. A 52% solution applied at low (40%) wet pickup, with magnesium chloride as the catalyst, imparted wrinkle recovery angles and DP appearance ratings about equal to those with DHDMI [65].

The reaction of equimolar amounts of glutaraldehyde and acrylamide in aqueous solution apparently yielded a 1:1 adduct. The product was formulated as 1-acryloyl-2,6-dihydroxypiperidine [67]. As a finishing agent, it produced higher wrinkle recovery angles than DHDMI, but the same DP ratings, in treated fabric, when magnesium chloride catalysis was used. Some fabric discoloration occurred.

F. Appraisal of Dialdehyde–Amide Adducts as Finishing Agents

The 1:1 adducts of dialdehydes with amides do not appear promising as DP finishing agents, in view of their tendency to produce discoloration of the fabric. The moderate DP performance imparted would not justify the added chemical costs involved. Moreover, acrylamide and its N-substituted derivatives present toxicity problems.

G. Adducts of Glyoxal Monoacetals with Ureas

Glyoxal can be reacted with low-molecular-weight alcohols to form monoacetals which can later be reacted with a linear or cyclic urea to yield DP finishing agents [68]:

$$OHC-CHO + 2ROH \rightarrow (RO)_2HC-CHO + H_2O$$

(I)

$$2(I) + R'NHCONHR' \rightarrow$$
$$(RO)_2CH-CHOH-NR'CONR'-CHOH-CH(OR)_2$$

(II)

The methylol agent **(II)** contains two acetal groups as additional reactive centers, making it theoretically capable of cross-linking four to six cellulose chains. An extra variation is to convert **(II)** to a ''capped'' DP finishing agent by alkylating the two N-methylol hydroxyls with a low-molecular-weight alcohol such as methanol in the presence of a mineral acid [69].

H. Appraisal of Glyoxal Monoacetal Adducts with Ureas

These agents appear to impart a medium level of permanent creasing as well as moderate wrinkle recovery angles, with tensile strength losses that are normal to cellulose cross-linking processes. Magnesium chloride–acetic acid was used as the curing catalyst. The adducts (**II**) and their "capped" derivatives can be isolated as pure compounds [69]. The use of agents (**I**) and (**II**) in DP finishing may produce alcohols as volatile coproducts, because the acetal groups in these molecules probably participate in the cellulose cross-linking process, as do also any "capped" *N*-methylol groups present.

VI. PHOSPHORIC OR PHOSPHONIC ACIDS AND THEIR SALTS

A. Processes Studied

Cotton cellulose has been partially converted to its phosphate esters by heating fibers, yarn, or fabric impregnated with mixtures of monosodium and disodium phosphate, or more effectively, with sodium hexametaphosphate. When the level of combined phosphorus in treated fabric exceeded 1.6%, the fibers were observed to be insoluble in cupriethylenediamine solution [70]. This is one indication that cross-linking had occurred in the cellulose:

$$2 \text{ cell—OH} + \text{NaH}_2\text{PO}_4 \rightarrow \text{cell—O—P(O)(ONa)—O—cell} + 2 \text{ H}_2\text{O}$$

Partial phosphorylation of cotton fabric has also been carried out by pad-dry-cure treatment with a solution containing monoammonium or diammonium phosphate together with urea, which served as a catalyst and fiber swelling agent. A moderate level of DP properties was imparted at sufficiently high levels of phosphorylation [71]. Polymer additives, such as cationic polyethylene together with an acrylic polymer of low T_g, were found to improve the DP performance and strength retention. On 50/50 cotton/polyester fabrics, extremely high wrinkle recovery angles (289–307°) and fair DP appearance ratings (3.5–3.7) resulted. High concentrations (e.g., 12% of monosodium or monoammonium phosphate and 12–24% concentrations of urea) were required. Curing temperatures of 160–170°C for relatively long times (4–7 min) were necessary and tended to cause fabric discoloration.

Cyanamide has been used as an impeller in the cross-linking of cotton with phosphoric acid to impart wrinkle resistance [72]. Urea has been used in this system as a catalyst [73]. Nitrilotris(methylenephosphonic acid) (NTMA) is another unusual cellulose cross-linking agent which has been applied to cotton fabric with cyanamide as the impeller or curing agent [74]. With 10% NTMA, 10%

cyanamide, and a fabric softener, pad-dry-cure treatment of cotton fabric imparted DP appearance ratings of 3.0. The tensile strength retention was 57%. Curing was at 163°C for 5 min. Although cyanamide is toxic, it is converted in acidic solution at a slightly elevated temperature to dicyandiamide and other less toxic agents.

B. Appraisal of Phosphorylation Processes as a Method of Finishing

The cross-linking of cotton through phosphorylation requires somewhat longer curing times than do conventional DP finishing treatments. The levels of DP performance imparted by pad-dry-cure application of monoammonium phosphate and urea have been moderate on 100% cotton fabrics, but reached practical and useful levels on 50/50 cotton/polyester. Notably, this type of process uses low-cost chemicals that are not mutagenic. The development of more effective impellers or coreactive catalysts could improve the prospects for such finishes. The possibility of obtaining launderable fabrics that are both wrinkle and flame resistant is an added incentive for further research and development. The durability of the flame resistance of such phosphorus-containing finishes depends greatly on the fixation of combined nitrogen as a synergist, in a form not removable by laundering. Methods for eliminating the ion-exchange properties of phosphorylated cotton are also needed, because the flame resistance of treated fabric decreases as sodium, magnesium, or calcium ions are taken up during laundering.

VII. POLYCARBOXYLIC ACIDS

A. Features of Ester-Type Cross-link Formation

Compounds having two or more carboxyl groups in each molecule are sometimes referred to as polycarboxylic acids. They are capable of forming cross-links in cotton by esterifying hydroxy groups of adjacent cellulose chains:

$$2 \text{ cell—OH} + \text{HOOC—R—COOH} \rightarrow \text{cell—OOC—R—COO—cell}$$

Tricarboxylic and tetracarboxylic acids have proven to be much more effective than dicarboxylic acids. Strong mineral acids are classical esterification catalysts, but they cannot be used in high-temperature textile finishing because they extensively degrade the cellulose. Instead, the polycarboxylic acid can furnish its own hydrogen ions as a type of autocatalysis.

In 1963, it was shown by Gagliardi and Shippee [75] that polycarboxylic acids are capable of imparting wrinkle resistance and shrinkage resistance to fabric of cotton, viscose rayon, and linen. Citric acid was the most effective agent tested, although it caused more fabric discoloration than other polycarboxylic acids during the long heat curing required. With 20% citric acid and curing at 143–163°C for 15–60 min, moderate to medium wrinkle recovery angles (250°–270°, warp + fill) were observed in treated cotton printcloth. The losses of tensile strength were very high, typically 61–65% in warp and filling after a 30-min cure at 160°C. The ester cross-links could be removed by saponification with hot $0.1N$ sodium hydroxide. This resulted in loss of wrinkle resistance but no restoration of the original strength. Evidently acid-catalyzed degradation of cellulose had taken place. Among other polycarboxylic acids tried was 1,1,4,4-butanetetracarboxylic acid, which proved less effective than citric acid.

Citric acid finishes were found to contain an abundance of unreacted carboxyl groups. These were shown to bind heavy metal cations such as copper, silver, and tin to the fabric to impart rot resistance [75]. The presence of free carboxyl groups also greatly increased the affinity of basic dyes such as Malachite Green for the fabric. Normally, the dye absorption of cotton fabrics is decreased by cross-linking treatments. Adsorption of cationic fabric softeners and water repellents was also increased by the free carboxyl groups in the citric acid finish [75].

B. Base Catalysis of Ester-Type Cross-linking

A step forward in polycarboxylic acid finishing was the use of alkaline catalysts such as sodium carbonate or triethylamine [76]. Enough of the base was added to neutralize 10–50% of the carboxyl groups. Acids having four to six carboxyl groups per molecule were usually much more effective than those with only two to three carboxyls. The true catalysts appeared to be monosodium or monoamine salts of these polycarboxylic acids. The salts also acted as efficient buffers and greatly decreased acid-induced tendering during the heat cure, which was at 160°C for 10 min. For wrinkle recovery angles of 248°–272°, breaking strength losses were 21–38% (warp direction). Further evidence of cross-linking was the insolubility of treated fibers in $0.5M$ cupriethylenediamine solution.

Surprisingly, these finishes were recurable. Creases durable to laundering could be introduced by simply ironing them into the rewet, cross-linked cotton fabric [77,78]. Ester cross-links appeared to be mobile at high temperature. It is probable that transesterification of ester groups is a pathway by which existing cross-links can be broken and new ones formed in the process of ironing the creases into the fabric. When cotton fabric cross-linked with all-*cis* 1,2,3,4-cyclopentanetetracarboxylic acid (CPTA) was recured with ethylene glycol present, the ester content decreased by 40%, causing a 65° drop in wrinkle recovery angle and loss of a launderable crease that had previously been set in place by cross-linking. Ethylene glycol had transesterified the cellulose ester, thus breaking the cross-links.

Concerning the mechanism of thermally-induced transesterification, it was found that heating alkyl hydrogen phthalates produced phthalic anhydride. If isoamyl alcohol were also present, the transesterification product as well as phthalic anhydride were obtained [77]. These results with model compounds led to the proposal that at high temperature, cellulose acid ester cross-links redissociate for form cyclic anhydrides of the polycarboxylic acid from which the cross-links were originally formed. The cyclic anyhydrides then esterify neighboring hydroxyl groups of the cellulose. The mobility of ester cross-links is due to reversible formation of cyclic anhydrides from cellulose acid esters. The particular acids effective as recurable finishing agents were those capable of forming five- or six-membered anhydride rings. Such polycarboxylic acids possess carboxyl groups attached to successive carbon atoms of a chain or ring. Examples are CPTA, already mentioned, and 1,2,3,4-butanetetracarboxylic acid (BTCA).

C. High-Speed Esterification Catalysts

A series of weak base catalysts have been discovered that are more active than sodium carbonate or tertiary amines. Alkali metal salts of phosphoric, polyphosphoric, phosphorous, and hypophosphorous acids have proven effective [79–84]. In order of decreasing activity as catalysts for BTCA finishing of cotton, they are ranked as follows:

$$NaH_2PO_2 > Na_2HPO_3 = NaH_2PO_3 > NaH_2PO_4 > Na_2H_2P_2O_7 > Na_4P_2O_7$$
$$> Na_5P_3O_{10} = (NaPO_3)_6 > Na_2HPO_4 = Na_3PO_4 > Na_2CO_3$$

These comparisons are based on DP appearance ratings imparted by 6.3% BTCA in 90-s cures at 180°C and on the durability of the resulting finishes to repeated laundering at a pH of 10. The laundering durability was measured in terms of the number of launderings and tumble dryings through which the DP appearance ratings remained at or above 3.5.

Sodium hypophosphite is the most effective catalyst and affords the most satisfactory whiteness in treated fabric. The agent acts as a reductive bleach in acid solution, is very weakly alkaline, and does not furnish hydrogen ions. Unfortunately, sodium hypophosphite is also one of the more expensive catalysts. The amount needed can be decreased by using the other catalysts as extenders. Monosodium phosphate, disodium pyrophosphate, or a mixture of tetrasodium pyrophosphate with phosphoric acid can be added for this purpose [84].

A number of polycarboxylic acids have been compared for effectiveness as DP finishing agents for cotton fabric, with sodium hypophosphite as the curing catalyst [80,83]. The pad-dry-cure treatments were carried out with 1% polyethylene softening agent present. Curing was at 180°C for 90 s in most cases, although other time–temperature combinations have also been used, such as 215°C for 15 s [84]. DP ratings of 4.3–4.7, were imparted, as well as conditioned wrinkle recovery angles of 295°–300°. The laundering durability of the finishes depended critically on the particular polycarboxylic acid used. The acids listed in order of decreasing laundering durability, together with the number of home launderings withstood with a DP rating ≥ 3.5, were as follows: BTCA (96) > mellitic (66) = CPTA (63) = 1,2,3-propanetetracarboxylic (68) > thiosuccinic (40) > citric (31) $\ggg$ maleic (5) > succinic (0). Although the number of carboxyl groups per molecule of finishing agent was important, other factors, such as substituent effects and molecular size, also helped determine the extent of cross-linking, as well as the rate of alkaline hydrolysis of the finish during subsequent laundering.

The breaking strength retention in polycarboxylic acid finishing was typically 54–57% and the tearing strength retained was 60–68% on cotton printcloth, as compared with 44% breaking strength and 54% tearing strength retention using DMDHEU catalyzed with magnesium chloride [83]. The latter finishing formulation imparted a DP rating of 4.6 and a wrinkle recovery angle of 303°, warp + fill. A fabric softener was used in these treatments. The polycarboxylic acid finishes consistently afforded appreciably improved breaking strength retention, due to the buffering action of the hypophosphite catalyst and the absence of Lewis-acid catalysts, which can cause cellulose chain cleavage.

D. Finishing with 1,2,3,4-Butanetetracarboxylic Acid

The commercial preparation of BTCA is a two-step synthesis. The Diels–Alder addition of butadiene to maleic anhydride yields 1,2,3,6-tetrahydrophthalic anhydride. This is oxidized to cleave the carbon–carbon double bond and form BTCA [85,86]. Currently produced on a pilot-plant scale, BTCA is, at present, too expensive to use as a DP finishing agent. The projected price of BTCA for very large-scale production would be about double the present price of DMDHEU.

1,2,3,4-Butanetetracarboxylic acid is the most effective of the polycarboxylic acids which have been studied thus far, in terms of DP performance imparted, speed of curing, laundering durability of the finish, and whiteness of treated fabric. The agent is normally applied at 4.0–5.5% levels, based on weight of fabric. A detailed and comprehensive comparison between BTCA and DMDHEU has been made with respect to DP appearance rating, tensile and tearing strengths, shrinkage resistance, and flex abrasion resistance in treated fabric. These properties were measured initially and after 30 launderings [87]. With sodium hypophosphite as the catalyst, the DP performance of BTCA was equal to that of DMDHEU catalyzed by magnesium chloride. BTCA finishing afforded a higher retention of tensile strength, tearing strength, and flex abrasion resistance. The shrinkage resistance with BTCA was moderately less than for DMDHEU. The two agents were also compared for any tendency to cause color shade changes in dyed fabrics. With most dyes, the agent gave similar results. Toward some sulfur dyes, however, the sodium hypophosphite used as a catalyst in BTCA finishing acted as a reducing agent. As nonreducing catalysts, disodium phosphite [87] and potassium dihydrogen phosphate [88] have been recommended for fabrics dyed with sulfur dyes or turquoise reactive dyes.

1,2,3,4-Butanetetracarboxylic acid finishes contain unesterified carboxyl groups and readily take up basic dyes [83,87,89]. The increased dye absorption is directly proportional to the concentration of BTCA applied and can be used as a quantitative measure of the level of treatment [87]. The afterdyeing of polycarboxylic acid finishes with cationic dyes is of interest, as it offers a possible means of dyeing garments which have already been DP finished in fabric or garment form [89].

Increased affinity for anionic dyes can be imparted to cotton by BTCA finishes by including in the treating formulation a tertiary amine possessing one or more hydroxyl groups per molecule. The alkanolamine becomes bonded to the cotton cellulose by the cross-linking agent and furnishes cationic centers to which the anionic dye quickly becomes affixed [90].

Triethanolamine as an additive in BTCA–hypophosphite formulations can also serve as a cross-link modifier that enhances DP appearance ratings, laundering durability of the finish, and fabric strength retention [91]. With both malic acid and triethanolamine added, it was found possible to eliminate all loss of flex abrasion resistance [92]. At a level of treatment imparting DP appearance ratings of 3.8–4.5 and wrinkle recovery angles of 262°–270°, the flex abrasion resistance was 98–115% of that for untreated fabric and double that for DMDHEU-treated fabric of the same DP performance. The formulations contained 0.5% polyethylene as a fabric softener. The finishes were durable to 58–92 laundering cycles at a wash temperature of 60–64°C (140–147°F). The most favorable mole ratios of triethanolamine/(BTCA + malic acid) were 0.90–1.00 in formulations containing 6% BTCA, 1.8% malic acid, and 1–2% sodium hypophosphite.

E. Mechanism of Base-Catalyzed Cross-linking by Polycarboxylic Acids

It is known that polycarboxylic acids such as BTCA form cyclic anhydrides readily when heated to sufficiently high temperatures [93,94]. Five- and six-membered anhydride rings are theoretically the easiest to form. The products actually produced from BTCA are the monoanhydrides and dianhydrides having five-membered rings. Weak bases are known to be effective catalysts for the esterification of cellulose with anhydrides. It was therefore proposed that base-catalyzed cross-linking of cotton by BTCA, CPTA, or other acids having carboxyl groups on adjacent carbon atoms of a molecular chain or ring proceeds via formation of cyclic anhydrides as the cellulose esterifying agents [83]. The base catalyst may increase the rate of cyclic anhydride formation, as well as increasing the rate of cellulose esterification and cross-linking by the anhydride.

According to the above theory, acids having only two carboxyls per molecule cannot cross-link cotton by base catalysis, because a pair of carboxyls is needed to form an anhydride ring. After esterification of one of the carboxyl groups, the remaining carboxyl is unable to form an anhydride ring, and so cannot esterify a second cellulose molecule to complete the cross-link. In agreement with this theory, only acids having at least three or four carboxyls per molecule are found to be effective cross-linking agents when using base catalysis. Exceptions are maleic acid and itaconic acid which can be polymerized or copolymerized *in situ* to higher polycarboxylic acids [95,96].

Thermogravimetric studies have shown that when BTCA or CTPA are heated, the loss of weight corresponds to water lost as anhydrides are formed. The mass spectra were characteristic of molecular fragments corresponding to cyclic anhydride formation [97,98]. The use of Fourier transform infrared (FTIR) diffuse reflectance spectroscopy has identified five-membered anhydride rings as the cellulose-reactive functional groups formed on heating polycarboxylic acids with cotton fabric, with and without sodium hypophosphite catalyst present [99]. In that study, 16 polycarboxylic acids, as well as polymaleic acid of number-averaged molecular weight 800, were examined. Two peaks representing symmetric and asymmetric stretching modes of the anhydride carbonyl group were observed in every case. As in earlier studies [100], the ester carboxyl group was detected by FTIR, after unesterified carboxyls were converted to carboxylate anions by dilute alkali. The absorbance gives a measure of number of ester groups formed. The effects that recuring the finished fabric have in the extent of cross-linking can be followed also [100]. Methods for quantitative analysis of BTCA [101] and citric acid [102,103] finishes on cotton fabric have been developed.

F. Phosphorus-Free Weak-Base Catalysts

When waste solutions of phosphates or other phosphorus-containing compounds are discharged into lakes or streams, such compounds are likely to serve as nutrients promoting the growth of algae. The algae lower the quality of the water for drinking and use up dissolved oxygen on which fish depend. Several types of phosphorus-free catalysts for BTCA finishing have therefore been developed.

Sodium salts of α-hydroxy acids such as malic (hydroxysuccinic), tartaric, and citric acids were rather effective catalysts in terms of initial DP appearance ratings (3.9–4.5) imparted by the BTCA and the number of launderings (89–122) the finishes could withstand. Wrinkle recovery angles of 266°–276°, warp + fill, were observed with a fabric softener present. Small amounts of boric acid were needed to prevent a slight yellowing of fabric during the heat cure at 180°C for 90 s. The catalyst/BTCA mole ratio was 1:2. The breaking strength retained was 60–65%; the tearing strength retention was 57–63% [104].

Monosodium and disodium salts of unsaturated dibasic acids such as maleic, fumaric, and itaconic acids exhibited greater catalytic activity in BTCA finishing of cotton than sodium salts of saturated acids such as formic, oxalic, malonic, glutaric, and adipic acids [105]. The whiteness of BTCA-treated fabrics was greater with monosodium and disodium fumarate as the curing catalyst than with monosodium or disodium salts of maleic or itaconic acid [106]. These studies were carried out using mercerized cotton printcloth. The breaking strength retentions were 65–75%, the tearing strength retained was 96–115%, and the flex abrasion resistance retained was 49–98% for maleate and fumarate catalysis. A nonionic polyethylene was used as a fabric softener. The DP appearance ratings were 3.2–4.0, and wrinkle recovery angles of 257°–281°, warp + fill, were observed, the values depending in some cases on the catalyst/BTCA mole ratio. The breaking strength and flex abrasion resistance retained were higher than with sodium hypophosphite catalysis in curing at 180°C for 90 s.

The sodium salts of chloroacetic acid, dichloroacetic acid, and trichloroacetic acid have been shown to catalyze BTCA cross-linking and DP finishing of cotton fabric [107]. Sodium chloroacetate was the most suitable, overall, of these three catalysts. With 6.3% BTCA, a DP rating of 4.0–4.3 was imparted, as well as a 261° wrinkle recovery angle. No fabric softener was used. The addition of boric acid or sodium tetraborate to the formulation improved the fabric whiteness. The breaking strengths of treated fabrics appeared about normal for cellulose cross-linking treatments. By comparison, sodium acetate was a relatively poor catalyst, its use resulting in a DP rating of 3.3 and a wrinkle recovery angle of 239°. Sodium hypophosphite as a catalyst led to a DP appearance rating of 4.5 and a wrinkle recovery angle of 259°, as well as a fabric whiteness index of 74 (untreated fabric = 76). However, sodium hypophosphite caused large shade changes in fabric dyed with three sulfur dyes studied. By contrast, sodium chloroacetate as a catalyst caused little changes in dye shade.

An entirely different type of catalyst is imidazole and its N-alkyl or C-substituted derivatives [108]. When imidazole was present with BTCA in mole ratios of 1:1 to 2:1, DP appearance ratings of 3.8–3.9 were imparted to mercerized cotton fabric. Wrinkle recovery angles of 250°–275° were observed. The tearing strength retention was 107–116%; the breaking strength retained was 71–77%. Flex abrasion resistance was 73–81% of that for untreated mercerized fabric. A fabric softener was present. A degree of fabric yellowing was evident, although much of this was removable by rinsing. N-Methylimidazole as the catalyst imparted less yellowing initially, but the tearing strength losses were moderately increased. Either catalyst led to higher retentions of strengths and flex abrasion resistance than sodium hypophosphite. The catalysis mechanism proposed was the reaction of BTCA to form cyclic anhydrides, which could then react with the imidazole-type catalyst to form acylimidazolium cations. The latter would esterify and cross-link cellulose. A practical advantage of such catalysts is that they caused little or no shade change in the kinds of dyes that are reduced by sodium hypophosphite [109].

G. Citric Acid Finishing

Citric acid is familiar to the cotton textile industry as a widely available, low-cost chelating agent and catalyst activator. This agent has been approved as safe for use in food and beverages and offers few environmental concerns.

Several studies have shown citric acid considerably less effective and less durable than BTCA as a DP finishing agent [80,83,110]. The α-hydroxyl group of citric acid interferes to some extent with the desired esterification and cross-linking of cellulose by this agent. The intereference was shown by direct comparison of citric acid and 1,2,3-propanetetracarboxylic acid (PCA) in the finishing of cotton printcloth [111]. The latter agent imparted a DP appearance rating of 4.6, a wrinkle recovery angle of 285°, warp + fill, and the finish was durable to over 70 launderings. Citric acid at the same molality (0.36) imparted a DP rating of 4.2 and a wrinkle recovery angle of 278°, and the finish was durable to 22 launderings. If 1% PCA was added to 7% citric acid formulation, the DP rating increased to 4.7 and the finish was durable to over 80 launderings. The only difference between citric acid and PCA is the α-hydroxyl group in citric acid molecules. In these treatments, the catalyst was 4.8% sodium hypophosphite monohydrate, curing was at 180°C for 90 s, and 0.5% polyethylene was the fabric softener used.

The adverse effect of the α-hydroxyl group of citric acid on the esterification and cross-linking of cellulose has been measured by FTIR spectroscopy [112]. The PCA-treated cotton showed higher absorbance by the ester carbonyl band than did citric-acid-treated cotton, over a wide range of curing temperatures. This correlated well with the greater wrinkle recovery angles imparted by PCA than by citric acid over the same range of curing temperatures.

A detailed study of DP finishing with citric acid in the presence of monosodium or disodium phosphate, pyrophosphate, and hypophosphite catalysts has been made [113]. Sodium hypophosphite was the most effective curing agent. With 5–7% citric acid and various concentrations of hypophosphite, the treatments imparted DP ratings of 3.5–4.0, conditioned wrinkle recovery angles of 247°–268°, and warp breaking strength retentions of 55–61%. Curing was preferably at 170–180°C for 60–90 s. The degree of fabric whiteness in the most favorable cases was comparable to that obtained in DMDHEU finishing. With regard to DP performance, fabric breaking strength, and fabric whiteness, the treatments were found to be an improvement over DMDMI finishing which used magnesium chloride catalyst activated with citric acid.

Selecting the citric acid concentration, catalyst concentration, cure temperature, and cure time that gave the best compromise between DP performance and fabric whiteness was difficult. By contrast, if a 6% concentration of a 3:1 mixture by weight of citric acid and BTCA were used as the cross-linking agent with a 6.2% hypophosphite catalyst, the whiteness index reached or surpassed that afforded by DMDHEU finishing, with DP ratings of 4.0–4.3 for a number of curing temperature–time combinations [113]. The citric acid was serving as an inexpensive extender for BTCA. Citric acid has also been used as an extender for cyclopentanetetracarboxylic acid at a weight ratio of 2:1. In this instance, the curing catalyst was monosodium phosphate [79].

Other curing additives which improve the fabric whiteness obtained in citric acid finishing of cotton include triethanolamine and several of its salts [114,115], triisopropanolamine, N-methyldiethanolamine, and polyethylene glycols [115] of 2–12 monomer units. Glycerol, pentaerythritol, sorbitol, and other polyhydric alcohols are also effective [115]. Boric acid and borates suppress yellowing [115,116] but can also lower the DP performance [115]. The whiteness index of fabric just removed from the curing oven increases as the fabric regains moisture from the air. Additives which serve as humectants often improve the whiteness. It is probable that moisture causes partial reversal of dehydration reactions responsible for the yellowing produced by heat curing [115].

H.   Reactive Activators for α-Hydroxy Acids

Mixtures of 1.5% BTCA and 4.5% citric acid imparted smooth-drying properties intermediate between those produced by 6% BTCA or 6% citric acid, in the presence of sodium hypophosphite as the catalyst [113]. Recently, it has been found that with a higher citric acid concentration (7%), the addition of as little as 0.25–0.50% BTCA causes unexpectably large increases in DP performance and laundering life of the resulting finish [111]. To account for this effect, it was proposed that a molecule of BTCA can esterify the α-hydroxyl group of one to three molecules of citric acid to produce oligomeric polycarboxylic acids, which may then esterify the α-hydroxyl groups of still other citric acid molecules. The resulting polymeric polycarboxylic acids can serve as the actual cellulose cross-linking agents. DP appearance ratings of 4.6–4.7 and wrinkle recovery angles of 278°–284°, warp + filling, were observed with a fabric softener present. The finishes were durable to 58–88 launderings, or 2–4 times as many as without BTCA added. Sodium hypophosphite was used as the curing catalyst.

Other activators for citric acid finishing are tartaric, maleic and phosphoric acids, as well as 1-hydroxyethane-1,1-diphosphonic acid [111]. More than one mechanism may be involved in their action, because they are fairly strong acids and may accelerate the esterification and cross-linking of cotton by classical hydrogen-ion catalysis. The first and last of these four additives are hydroxy acids and might be esterified by citric acid to give cross-linking agents of increased functionality. Maleic acid should esterify α-hydroxyl groups of citric acid in the same manner as BTCA or PCA. The action of small amounts of PCA in activating a high concentration of citric acid was noted in the preceding section.

Physical evidence for the esterification of the α-hydroxyl group of citric acid by added polycarboxylic acids is provided by recent FTIR spectral data for reaction mixtures of polymaleic acid and citric acid [112]. The polymaleic acid had an average chain length of seven monomer units. Heating polymaleic acid at 180°C for 2 min caused cyclic anhydride groups to form. The quantity of anhydride groups greatly decreased when citric acid was present during the heating. The intensity of the anhydride carbonyl band decreased as the proportion of citric acid in the original mixture was increased. Analogous results were obtained on heating polymaleic acid with other hydroxy acids such as tartaric acid or malic acid. It was concluded that the reaction products are more reactive toward cellulose because (1) they have more pairs of carboxyl groups for cyclic anhydride formation and (2) they are free of interfering α-hydroxyl groups, as a result of esterification of the latter by the polymaleic acid.

I. Malic Acid as a DP Finishing Agent

The molecular structure $HO-CH(COOH)CH_2COOH$ for malic acid (hydroxysuccinic acid) makes this compound seem an unlikely cross-linking agent, because three carboxyls per molecule are needed for effectiveness in the cyclic anhydride mechanism of cellulose cross-linking. However, if BTCA is also present during the heat cure, each molecule of BTCA can esterify the α-hydroxyl group of one to three molecules of malic acid to form an oligomeric acid having several pairs of carboxyls available to cross-link cotton. DP appearance ratings of 4.3–4.4 were readily imparted to cotton printcloth by 5.4% malic acid activated with 2–3% BTCA [111]. The catalyst was 3.2% sodium hypophosphite monohydrate, curing was at 180°C for 90 s, and 0.5% low-molecular-weight polyethylene was present as a fabric softener. The conditioned wrinkle recovery angles of 272°–273° were observed. The breaking strength retention was 59–64% and the tearing strength retention was 52–55%, measured in the warp direction. The whiteness index was comparable to that in DMDHEU finishing and was considerably higher for malic acid than for citric acid treatments. This difference in whiteness was attributed to a difference in the thermal dehydration products possible from these two acids.

In later studies [117], the BTCA concentration was decreased to 1% in the malic acid formulation while increasing the catalyst concentration to 4.8%, adding 1.5% H_3PO_4, and using a high-density polyethylene as a fabric softener. The DP rating imparted was 4.3, the wrinkle recovery angle was 271°, the breaking strength retention was 68%, and tearing strength retention was 64%. The finish was durable to over 70 laundering cycles. Further development of malic acid finishing is likely.

J. Maleic Acid and Its Polymers as Finishing Agents

Having the structure HOOCCH $=$ CHCOOH which contains only two carboxyl groups per molecule, maleic acid imparts only moderate to fair DP properties even in the presence of phosphoric acid plus sodium hypophosphite as the mixed catalyst [111]. However, it is possible to form copolymers of maleic acid *in situ* in the fibers of cotton fabric in the presence of free-radical initiators, as the fabric is dried, prior to curing. Sodium hypophosphite can be included as a catalyst for subsequent high-temperature cross-linking of the cellulose. Such a process has been carried out with maleic acid and itaconic acid as comonomers [95,96]. Without a comonomer present, maleic acid does not polymerize.

On cotton printcloth, DP appearance ratings of 4.0–4.4 and wrinkle recovery angles of 268°–283° were imparted by fabric treatment with 12% of an equimolar mixture of maleic acid and itaconic acid (methylenesuccinic acid), 0.18% potassium persulfate, 8.8% sodium hypophosphite monohydrate, and polyethylene as the fabric softener. Copolymerization took place during drying at 100°C for 10 min. The copolymer cross-linked to cotton cellulose during the cure at 160–190°C for 2.0–3.5 min [95].

An extraordinary feature of this type of finish was the high flex abrasion resistance retained (143–214% of the valued given by untreated fabric). For a DMDHEU finish, the flex abrasion resistance retained was only 18%. The whiteness index for the copolymer finishes was comparable to that for DMDHEU treatment. The warp tearing strength retention (51–83%) and breaking strength retention (49–60%) depended on the conditions used. These copolymer finishes caused a decrease in measured stiffness of the cotton fabric.

Treatment of mercerized printcloth with 9% comonomer mixtures in the above type of formulation, with persulfate initiator, 6.4% hypophosphite catalyst, and fabric softener, was carried out with drying as usual and curing at 170°C for 1.5 min. The DP rating was 4.0, the wrinkle recovery angle was 260°, the tearing strength retention was 77%, the breaking strength retention was 62%, and the flex abrasion resistance retained was 93% [96]. When the predrying step was carried out at a temperature as high 140°C for 10 min, the final curing step could be omitted. This might be especially attractive for garment curing, where multiple layers of fabric normally require slow, even heating to obtain uniformity of cross-linking.

Instead of copolymerizing maleic acid and other vinyl or acrylic agents *in situ* in the fibers of cotton fabric, it has been found possible to apply preformed homopolymers and copolymers of maleic acid. These are prepared commercially by homopolymerization and copolymerization of maleic and anhydride, followed by hydrolysis of the anyhydride groups. Such polymeric acids have a high degree of solubility in water. The degree to which they penetrate into cotton fibers decreases as their molecular weight and size are increased. The lowest homopolymer of maleic acid commercially available has a number-averaged molecular weight of about 800. This corresponds to approximately seven monomer units per molecule. This heptameric acid has been applied to printcloth at a concentration of 8% together with 3–6% sodium hypophosphite and curing at 180–190°C for 2–3 min. A terpolymer of maleic acid also was tried under these conditions. In the absence of a fabric softener, the wrinkle recovery angles imparted were 232°–260°, warp + filling. With 6% BTCA as the finishing agent, the wrinkle recovery angle was 286°. The terpolymer of maleic acid was more effective than the homopolymer [112]. Finishing of cotton with either the homopolymer or terpolymer, together with 2–8% citric acid as an extender, was also monitored by FTIR spectroscopy. The anhydride carbonyl band absorbance in treated fabric decreased as the citric acid concentration was raised in the treating formulation. This showed that the cyclic anhydride formed at a high temperature by polymaleic acids had esterified α-hydroxyl groups of citric acid molecules to form still higher polycarboxylic acids as the actual finishing agents.

Analogous DP finishes with proprietary terpolymer "polycarboxylic blends" have been reported to impart DP appearance ratings of 4.0–4.3, wrinkle recovery angles of 292°–297°, and improved retention of strength and abrasion resistance, relative to DMDHEU finishing. The curing catalyst was 3–5% sodium hypophosphite monohydrate, and the fabric softener was polyethylene [118]. Curing was at 185–190°C for 2–3 min.

The advantage of using an α-hydroxyl acid such as citric acid in conjunction with polymaleic acid in DP finishing would lie in the ability of citric acid to cross-link cellulose in the interior of cotton fibers where the polymaleic acid cannot penetrate well because of its large molecular size. The polymaleic acid grafted in the outer layers of each fiber would undoubtedly be coupled via ester linkages with α-hydroxyl groups of citrate groups a little further in the interior, thus providing bonding between the interior and exterior finishes.

K. Appraisal of Polycarboxylic Acids as DP Finishing Agents

The most effective agent of this class is BTCA, which may be regarded as a dimer of maleic acid, and is the first member of the polymaleic acid homologous series. The heptamer of maleic acid is commercially available, is far less expensive than BTCA, and is undergoing commercial development as a finishing agent for cotton textile. Fairly low-molecular-weight copolymers and terpolymers of maleic acid are also under development for this purpose. If oligomers intermediate between the dimer and heptamer of maleic acid could be made at low cost, these would offer even greater promise because of the much greater accessibility of the internal regions of cotton fibers to small-sized molecules than to large ones. Formaldehyde-free cross-linking agents cause less collapse of internal pore structure of the cotton fiber than does DMDHEU, at comparable levels of resilience imparted [119], and this is accompanied by improved retention of breaking strength, tearing strength, and flex abrasion resistance. The choice of catalyst used likewise affects the residual pore volume of the treated fibers and, in turn, the breaking strength and flex abrasion resistance retained [120]. Sodium hypophosphite, the most effective catalyst for polycarboxylic acid finishing, is also the most expensive. Disodium phosphite, the next most effective catalyst, has the advantage that it does not cause color shade changes in treating fabrics dyed with sulfur dyes [87].

Citric acid is an economical, environmentally acceptable cellulose cross-linking agent. In the presence of 0.5% BTCA together with 1.5% phosphoric acid as the coactivator, and sodium hypophosphite or disodium phosphite as the catalyst, citric acid imparts a high level of DP properties, and the finish withstands home laundering detergents well at wash temperatures of 50°–65°C. Citric acid finishes tend to cause a faint discoloration in white fabric, except at very high concentrations of hypophosphite, and are best suited for dyed materials. Citric acid can also be used as an extender for maleic acid polymers in DP finishing. Malic acid (hydroxysuccinic acid) is similar to citric acid as a low-cost, environmentally innocuous agent. Unlike citric acid, it is nonyellowing during heat curing, but in the absence of 0.5–1.0% BTCA as an activator, it is almost completely ineffective as a cross-linking agent.

Niche uses for polycarboxylic acid finishing have appeared in the treatment of wood pulp and paper. Disposable diapers containing kraft pulp fibers crosslinked with citric acid and sodium hypophosphite have been widely commercialized. The fibers so treated have increased moisture absorbency as well as increased resiliency when wet or dry. The formaldehyde-free nature of the cross-linking agent is particularly important for this type of application [121,122]. The application of polymaleic acid heptamer and sodium hypophosphite to kraft paper has been shown to increase the wet strength of the paper [123]. The wet strength was directly proportional to the ester carbonyl band absorbance measured by FTIR diffuse reflectance spectroscopy. The improvement of wet strength was due to ester-type grafting or cross-linking of the cellulose by the polymaleic acid. In identifying polycarboxylic acid finishes applied to various types of cellulose, FTIR spectroscopy is of particular value in detecting ester linkages and also in identifying the type of catalyst used [124], because appreciable amounts of curing catalyst remain in the treated materials [79,113]. The presence of small amounts of combined phosphorus in the trivalent state can also be detected by electron spectroscopy for chemical analysis [124].

Polycarboxylic acid DP finishing has also been applied to silk fabrics [1]. Fibroin, the silk protein, is built up of peptide units derived from a variety of amino acids, including serine, threonine, and tyrosine, which contain hydroxyl groups with which BTCA can cross-link. An extraordinary feature of the resulting

DP silk fabrics was their extremely high tearing strength retention, exceeding 200% in warp and filling at wrinkle recovery angles of 300° or higher! The breaking strength was over 90% of that for untreated fabric. This raises the possibility that fabric strength losses might not be inherent to all cross-linking processes in cotton, if the cotton fibers were suitably modified prior to the DP finishing treatment.

Polycarboxylic acid finishing, like formaldehyde-free DP finishing in general, is characterized by a great diversity of cross-linking agents and combinations of agents, an ongoing advance by university, governmental, and industrial laboratories in exploration, development, and adaption to existing commercial needs, and the recognition of new applications as economical and effective processes begin to emerge.

REFERENCES

1. Y. Yang and S. Li. Text. Chem. Color. *26*(5):25 (1994).
2. P. Wilcox, A. Naidoo, D. J. Wedd, and D. G. Gatehouse. Mutagenesis *5*:285 (1990).
3. J. A. Zijlstra. Mutat. Res. *210*:255 (1989).
4. H. Petersen. Rev. Prog. Color. *17*:7 (1987).
5. T. M. Monticello, K. T. Morgan, J. L. Everitt, and J. A. Popp. Am. J. Pathol. *134*: 515 (1989).
6. Assessment of health risks of garment workers and certain home residents from exposure to formaldehyde, Executive Summary XIII–XV, and Fact Sheet, U.S. Environmental Protection Agency, April 1987.
7. H. Petersen and N. Petri. Melliand Textilber. *66*:285 (1985).
8. F. Reinart. Textilveredlung *24*:223 (1989).
9. B. North. Text. Chem. Color. *23*(4):23 (1991).
10. B. North. Text. Chem. Color. *23*(10):21 (1991).
11. P. J. O'Brien, H. Kaul, L. McGirr, D. Drolet, and J. M. Silva. Pharmacol. Eff. Lipids *3*:266 (1989).
12. R. M. Reinhardt and B. A. K. Andrews. Text Chem. Color. *16*(11):29 (1984).
13. J. D. Turner, Durable Press Garments, brochure by Cotton Incorporated, 1994.
14. F. S. H. Head. J. Text. Inst. *49*:T345 (1958).
15. E. J. Gonzales and J. D. Guthrie. Am. Dyest. Rep. *58*(3);27 (1969).
16. K. Yamamoto. Text. Res. J. *52*:357 (1982).
17. C. M. Welch and G. F. Danna. Text. Res. J. *52*:149 (1982).
18. M. D. Hurwitz and L. E. Condon. Text. Res. J. *28*:257 (1958).
19. C. M. Welch. Text. Res. J. *53*:181 (1983).
20. C. M. Welch and J. G. Peters. Text. Res. J. *57*:351 (1987).
21. C. M. Welch. Text. Chem. Color. *16*:265 (1984).
22. D. L. Worth, U.S. Patent 4,269,603 to Riegel Textile Corp. (1981).
23. D. L. Worth, U.S. Patent 4,269,602 to Riegel Textile Corp. (1981).
24. Cosmetic Ingredient Review. J. Am. Coll. Toxicol. *14*:348 (1995).
25. J. G. Frick, Jr. and R. J. Harper, Jr. J. Appl. Polym. Sci. *27*:983 (1982).

26. J. G. Frick, Jr. and R. J. Harper, Jr. J. Appl. Polym. Sci. *28*:3875 (1983).
27. E. B. Whipple and M. Ruta. J. Org. Chem. *39*:1666 (1974).
28. J. P. Shyu and C. C. Chen. Text. Res. J. *62*:469 (1992).
29. R. Jung, G. Engelhart, B. Herbolt, R. Jaeckh, and W. Mueller. Mutat. Res. *278*: 265 (1992).
30. H. Matsuda, Y. Ose, T. Sato, H. Nagase, H. Kito, and K. Sumida. Sci. Total Environ. *117–118*:521 (1992).
31. S. Goto, O. Endo, T. Mizoguchi, H. Matsushita, T. Kobayashi, F. Fukai, T. Katayama, and Y. Oda. Kankyo Kagaku *3*:482 (1993); Chem. Abstr. *119*:243347y (1993).
32. D. Zissu, F. Gagnaire, and P. Bonnet. Toxicol. Lett. *72*:53 (1994).
33. E. A. Gross, P. W. Mellick, F. W. Kari, F. J. Miller, and K. T. Morgan. Fundam. Appl. Toxicol. *23*:348 (1994).
34. R. O. Beauchamp, Jr, M. B. St. Clair, T. R. Fennell, D. O. Clarke, K. T. Morgan, and F. W. Kari. Crit. Rev. Toxicol. *22*(3–4):143 (1992).
35. S. W. Frantz, J. L. Beskitt, M. J. Tallant, J. W. Futtrell, and B. Ballantyne. J. Toxicol. Cutan. Ocul. Toxicol. *12*:355 (1993).
36. G. B. Leslie. Indoor Built Environ. *5*(3):132 (1996).
37. J. G. Frick, Jr. and R. J. Harper, Jr. J. Appl. Polymer Sci *29*:1433 (1984).
38. G. Rotta, S. Wittman, and W. W. Volz, German Offen. DE 3,832,089 to Chemische Fabrik Theodor Rotta GmbH and Co. K-g. (1990).
39. L. H. Chance, G. F. Danna, and B. K. Andrews, U. S. Patent 4,900,324 to U.S. Dept. of Agriculture (1990).
40. L. H. Chance and G. F. Danna, U.S. Patent 4,818,243 to U.S. Dept. of Agriculture (1989).
41. A. Blanc, D. Wilhelm, and B. Caltot. Melliand Textilber. *76*:E181,711 (1995).
42. S. L. Vail, P. J. Murphy, Jr., J. G. Frick, Jr., and J. D. Reid. Am. Dyest. Rep. *50*: 550 (1961).
43. S. L. Vail, R. H. Barker, and P. G. Mennitt. J. Org. Chem. *30*:2179 (1965).
44. S. L. Vail and P. J. Murphy, Jr., U.S. Patent 3,112,156 to U. S. Depart. of Agriculture (1963).
45. H. M. Ziifle, R. R. Benerito, and E. J. Gonzales. Text. Res. J. *38*:925 (1968).
46. H. Z. Jung, R. R. Benerito, E. J. Gonzales, and R. J. Berni. J. Appl. Polym. Sci. *13*:1949 (1969).
47. H. Petersen, P. Pai, F. Kippel, and F. Reinart, U.S. Patent 4,295,846 to BASF A.-G. (1981).
48. R. J. Harper, Jr. and J. G. Frick, Jr. Am. Dyest. Rep. *70*(9):46 (1981).
49. K. Yamamoto. Text. Res. J. *52*:363 (1982).
50. J. G. Frick, Jr. J. Appl. Polym. Sci. *30*:3467 (1985).
51. J. G. Frick, Jr. Text. Res. J. *56*:124 (1986).
52. J. G. Frick Jr., B. W. Jones, R. L. Stone, and M. D. Watson, U.S. Patent 4,619,668 to U.S. Dept. of Agriculture (1986).
53. E. J. Blanchard and R. M. Reinhardt. Am. Dyst. Rep. *78*(2):30 (1989).
54. R. M. Reinhardt and E. J. Blanchard. Colour. Annu. 29 (1989/1990).
55. N. R. Bertoniere and W. D. King. Text. Res. J. *59*:608 (1989).
56. Takagishi and T. Saka, European Patent Application 320,010 (1989).
57. Sumitomo Co., Japanese Patent 59,116,476 (1984).

58. M. T. Beachem, U.S. Patent 3,260,565 to American Cyanamid Co. (1966).
59. X. Kaestele, M. Bernheim, and E. Roessler, European Patent Application EP 330,979 to Chemische Fabrik Pfersee GmbH (1989).
60. J. G. Frick, Jr. and R. J. Harper, Jr. Text. Res. J. 53:660 (1983).
61. I. Y. Slonim, B. M. Arshava, F. G. Zhurina, and T. K. Kaliya. Vysokomol Soedin Ser. A 32:513 (1990); Chem. Abstr. 113:24558r (1990).
62. B. F. North, U.S. Patent 4,285,690 to Sun Chemical Corp. (1981).
63. W. C. Floyd and B. F. North, U.S. Patent 4,455,416 to Sun Chemical Corp. (1984).
64. J. G. Frick, Jr. and R. J. Harper, Jr. Ind. Eng. Chem. Product Res. Dev. 21(1):1 (1982).
65. J. G. Frick, Jr. and R. J. Harper, Jr. Text. Res. J. 52:141 (1982).
66. J. G. Frick, Jr. and R. J. Harper, Jr. Text. Res. J. 51:601 (1981).
67. J. G. Frick, Jr. and R. J. Harper, Jr. Text. Res. J. 53:758 (1983).
68. W. Didier, S. Loison, and A. Blanc, U.S. Patent 4,968,774 to Societe Francaise Hoechst (1990).
69. W. Didier, A. Gelabert, and A. Blanc, U.S. Patent 4,854,934 to Societe Francaise Hoechst (1989).
70. D. M. Dallagher. Am Dyest. Rep. 53:361 (1964).
71. E. J. Blanchard and R. J. Harper, Jr. J. Coated Fabrics 12(4):213 (1983).
72. American Cyanamid, Japanese Patent 82 47 973 (1982).
73. American Cyanamid, Japanese Patent 82 47 974 (1982).
74. J. D. Turner. Text. Res. J. 49:244 (1979).
75. D. D. Gagliardi and F. B. Shippee. Am Dyest. Rep. 52:300 (1963).
76. S. P. Rowland, C. M. Welch, M. A. F. Brannan, and D. M. Gallagher. Text. Res. J. 37:933 (1967).
77. S. P. Rowland and M. A. F. Brannan. Text. Res. J. 38:634 (1968).
78. S. P. Rowland, C. M. Welch, and M. A. F. Brennan, U.S. Patent 3,526,048 to U.S. Dept. of Agriculture (1970).
79. C. M. Welch. Text. Res. J. 58:480 (1988).
80. C. M. Welch and B. K. Andrews. U.S. Patent 4,820,307 to U.S. Dept. of Agriculture (1989).
81. C. M. Welch and B. K. Andrews. U.S. Patent 4,936,865 to U.S. Dept. of Agriculture (1990).
82. C. M. Welch and B. K. Andrews. U.S. Patent 4,975,209 to U.S. Dept. of Agriculture (1990).
83. C. M. Welch and B. K. Andrews. Text. Chem. Color. 21(2):13 (1989).
84. C. M. Welch. Text. Chem. Color. 22(5):13 (1990).
85. J. E. Franz, W. S. Knowles, and C. Osuch. J. Org. Chem. 30:4328 (1965).
86. J. W. Lynn and R. L. Roberts. J. Org. Chem. 26:4303 (1961).
87. G. L. Brodman. Text. Chem. Color. 22(11):13 (1990).
88. D. L. Brotherton, K. W. Fung, and A. L. Addison. AATCC 1989 Internat. Conf. and Exhib., Book of Papers, 1989, p. 170.
89. B. A. K. Andrews, E. J. Blanchard, and R. M. Reinhardt. Am. Dyest. Rep. 79(9): 48 (1990).
90. E. J. Blanchard, R. M. Reinhardt, and B. A. K. Andrews. Text. Chem. Color. 23(5): 25 (1991).

91. C. M. Welch. Text. Chem. Color. *23*(3):29 (1991).
92. C. M. Welch. Text. Chem. Color. *29*(2):21 (1997).
93. J. Aubry and E. Yax, German Patent 2,056,937 to Societe Chimique des Charbonnages (1971).
94. R. V. Volkenburgh, J. R. Olechowski, and G. C. Royston, U.S. Patent 3,194,816 to Copolymer Rubber and Chemical Corp. (1965).
95. H.-M. Choi. Text. Res. J. *62*:614 (1992).
96. H.-M. Choi and C. M. Welch. Am. Dyest. Rep. *83*(12):48 (1994).
97. B. J. Trask-Morrell, B. A. K. Andrews, and E. E. Graves. Text. Chem. Color. *22*(10):23 (1990).
98. B. J. Trask-Morrell and B. A. K. Andrews. J. Appl. Polym. Sci. *42*:511 (1991).
99. C. Q. Yang and X. Wang. Text. Res. J. *66*:595 (1996).
100. C. Q. Yang. Text. Res. J. *61*:298 (1991).
101. C. Q. Yang and G. D. Bakshi. Text. Res. J. *66*:377 (1996).
102. N. M. Morris, B. A. K. Andrews, and E. A. Catalano. Text. Chem. Color. *26*(2): 19 (1994).
103. N. M. Morris, E. A. Catalano, and B. A. K. Andrews. Cellulose *2*:31 (1995).
104. C. M. Welch and J. G. Peters. Text. Chem. Color. *25*(10):25 (1993).
105. H.-M. Choi and C. M. Welch. Text. Chem. Color. *26*(6):23 (1994).
106. H.-M. Choi, C. M. Welch, and N. M. Morris. Text. Res. J. *64*:501 (1994).
107. B. A. K. Andrews. Ind. Eng. Chem. Res. *35*:2395 (1996).
108. H.-M. Choi, C. M. Welch, and N. M. Morris. Text. Res. J. *63*:650 (1993).
109. H.-M. Choi, J. D. Li, R. D. Goodin, and T. D. Pratt. Am. Dyest. Rep. *83*(2):38 (1994).
110. B. A. K. Andrews, C. M. Welch, and B. J. Trask-Morrell. Am. Dyest. Rep. *78*(6): 15 (1989).
111. C. M. Welch and J. G. Peters. Text. Chem. Color. *29*(3):22 (1997).
112. C. Q. Yang, X. Wang, and I.-S. Kang. Text. Res. J. *67*:334 (1997).
113. B. A. K. Andrews. Text. Chem. Color. *22*(9):63 (1990).
114. B. A. K. Andrews, E. J. Blanchard, and R. M. Reinhardt. Text. Chem. Color. *25*(3): 52 (1993).
115. H.-M. Choi. Text. Chem. Color. *25*(5):19 (1993).
116. K.-W. Fung, K. H. Wong, and D. L. Brotherton, U.S. Patent 5,199,953 to Ortec, Inc. (1993).
117. C. M. Welch and J. G. Peters. Text. Chem. Color. *29*(10):33 (1997).
118. C. Q. Yang and K. Y. Chang. AATCC 1996 Internat. Conf. and Exhib., Book of Papers, 1996, p. 160.
119. N. R. Bertoniere and W. D. King. Text. Res. J. *62*:349 (1992).
120. N. R. Bertoniere, W. D. King, and C. M. Welch. Text. Res. J. *64*:247 (1994).
121. C. M. Herron and D. J. Cooper, U.S. Patent 5,137,537 to The Procter & Gamble Cellulose Company (1992).
122. J. T. Cook, P. A. Rodriguez, P. A. Graef, C. R. Bolstad, and W. L. Duncan, U.S. Patent 5,562,740 to The Procter & Gamble Company (1996).
123. C. Q. Yang, Y. Xu, and D. Wang. Ind. Eng. Chem. Res. *35*:4037 (1996).
124. B. A. K. Andrews, N. M. Morris, D. J. Donaldson, and C. M. Welch, U.S. Patent 5,221,285 to U.S. Dept. of Agriculture (1993).

2
Surface Characteristics of Polyester Fibers

YOU-LO HSIEH   Department of Textiles, University of California at Davis, Davis, California

I. INTRODUCTION

Polyester has been one of the most popular fibers, second to cotton as measured by production tonnage in recent years. The technical merits and commercial versatility of the fiber production system have led to successful product development and applications. Polyester fibers have many desirable properties, including relatively high tenacity, low creep, good resistance to strain and deformation, high glass transition temperature, and good resistance to acids and oxidizing agents. These physical, mechanical, and chemical attributes make polyester fibers excellent candidates not only for apparel and textile products but also for industrial and composite applications. In apparel, for instance, polyester fibers are versatile because of their receptivity to heat treatments (setting and texturing) and the ease of their blending with other fibers such as cotton, wool, and regenerated cellulosics. However, polyester fibers also possess certain characteristics which constrain their use. Polyester fibers retain little moisture and do not transport aqueous fluids. The hydrophobic nature of polyester fibers makes them difficult to dye (they require a carrier) and to finish in aqueous media. Their oleophilic nature attracts oily soils and leads to poor adhesion to rubber and plastics. Polyesters also have poor resistance to alkalis.

Polyesters are polymers that contain ester linkage groups along their main chains [1]. For fiber-forming polyesters, the polymer structures have to be linear in order for the long chains to be oriented and organized when spun into the fibrous forms and, at the same time, sufficiently rigid to possess desirable glass transition and melting temperatures. The International Standard Organization (ISO) defines polyester fibers as those from polymers based on a diol and a terephthalic acid. The Federal Trade Commission (FTC) defines them as those which contain at least 85% by weight of an ester of a substituted aromatic carboxylic acid, including but not restricted to substituted terephthalate units and para-substituted hydroxybenzoate units.

Polyethylene terephthalate (PET) is the most common fiber-forming polyester. The properties of PET fibers depend strongly on the development of the microstructures during the fiber-formation (spinning, drawing, heat setting) process where the transformation from melt to solid occurs. Therefore, the physical properties of PET fibers vary depending on their processing conditions and thermal history. The common melt spinning and drawing processes of commercial PET produce fibers with an oriented and semicrystalline structure and a typical density around 1.38 g/cm^3. This is between the nonoriented amorphous and the calculated crystalline densities of 1.335 g/cm^3 and 1.45 g/cm^3, respectively [2].

The main physical properties of polyester fibers that influence their surface characteristics are thermal and light properties. The former affects the melt adhesion properties, whereas the latter affects the appearance. The melting (T_m) and glass transition (T_g) temperatures of PET are around 250–266°C and 70–80°C, respectively, relatively high for thermal processing.

Among the commercially available fibers, PET fibers have the highest birefringence and refractive index ($n_{\rm II}$) values. The birefringence of a fiber is determined by the difference between the principal refractive indices or the refractive indices parallel ($n_{\rm II}$) and perpendicular ($n_{\rm I}$) to the fiber axis ($n_{\rm II} - n_{\rm I}$). Whereas the refractive index values of most fibers lie in the range of 1.5–1.6, the $n_{\rm II}$ values of PET fibers are much higher (~1.725), resulting in birefringence values that are several times higher than most other fibers. The magnitude of the birefringence for a fiber depends on the degree of symmetry and orientation of the molecules. The degree of symmetry of a molecule is determined by the chemical structure of the polymer, whereas the degree of orientation of a molecule is determined by the processing (spinning and drawing) variables. The high symmetry of the benzene ring structure in PET is responsible for its high $n_{\rm II}$ value. Chain orientation, on the other hand, depends on fiber processing and affects both $n_{\rm II}$ and $n_{\rm I}$ values of PET. Like other manufactured fibers, PET fibers have varied birefringence, depending on how they are spun and drawn. In general, the high birefringence and $n_{\rm II}$ of PET result in higher light reflection, causing the fibers to appear lighter, especially of dark colors. This is one of the reasons that the production of dark shades is a greater challenge for microdenier polyester than other ultrafine fibers.

In addition to PET, other polyester fibers have been developed and commercialized. In this chapter, polyesters other than PET are briefly discussed. Although the surface characteristics of the other polyesters are of scientific interest and technical importance, little has been published on this subject. Over the years, polyester has become synonymous with PET. As the majority of research related to the surface structure, modification, and characterization of polyester fibers is on PET, this chapter focuses mainly on PET fiber surfaces. In cases where either the approaches for surface modification or the methods of surface characterization are relevant to polyester surfaces, work related to PET films is cited as well.

Surface characteristics such as liquid wetting, liquid repellency, soil release, adhesion, friction, handling, light reflection, and static property are important to processing and performance properties of fibrous products. Performance requirements for polyester for textile end uses such as noncarrier dyeability, soil-release properties, and those for nonwovens and composites such as bondability and adhesion have resulted in the development of modified polyesters to improve the less desirable properties without affecting the positive attributes.

This chapter begins with discussion of the chemical structures of polyesters, surface structure and chemistry, surface energetics, and wetting contact angle, followed by surface effects during fiber formation or spinning (fiber size and cross-sectional shape) and the effects of drawing and heating. It then goes into the surface-modification approaches which alter the chemistry and/or morphology of polyester fiber surfaces. These methods for altering the surface structure of polyester fibers and fabrics include hydrolysis (alkaline and enzymatic), surface grafting, plasma, and excimer ultraviolet (UV) laser.

II. CHEMICAL STRUCTURE OF POLYESTERS

A. Polyethylene Terephthalate

Polyethylene terephthalate (PET) constitutes the majority of the polyester fibers produced and consumed globally. The concept for forming the polyesters from diols and diacids was first described by Carothers in 1931 and was developed into a patent on PET by Whinfield and Dickson in 1941. PET is a polycondensation product of ethylene glycol and terephthalic acid (or dimethyl terephthalate). Polymerization of PET proceeds a two-step process in which the first step is a reaction between a 2-to-1 ratio of ethylene glycol and terephthalic acid (or dimethyl terephthalate) that leads to the formation of bis(hydroxyethyl)terephthalate (BHET). In the second step, transesterification of BHET yields PET. The typical weight-averaged molecular weight of commercial PET fibers is 40,000 Da, equivalent to an average PET chain length consisting of 200 repeating units.

$$\left[\begin{array}{c} \overset{\text{C}}{\underset{\text{O}}{\|}} - \bigcirc - \overset{\text{C}}{\underset{\text{O}}{\|}} - \text{O} - \text{CH}_2\text{CH}_2 - \text{O} \end{array}\right]_n$$

Theoretically, PET should contain only hydroxyl chain ends. Practically, carboxylic groups have been observed because of degradation (hydrolysis, thermal oxidation) during the polymerization and melting processing. Because diethylene glycol (DEG) forms as a side reaction by-product of the glycol, about 1–3% of DEG is present in PET as a statistical copolymer. The ether linkage in DEG increases the sensitivity of PET to photodegradation and thermal oxidation.

B. Other Polyesters [3]

Changing either the diol or the diacid structure in the PET structure creates new polyester structures with altered physical, chemical, and morphological properties. The changes in the chemical structures of either the diol or the diacid are expected to influence crystallization and the formation and stabilization of the microstructure in the polymer solids. Molecular structure, including chain stiffness and intermolecular cohesive forces, alters the crystallization behavior, the crystal structure, and overall morphology, thus the thermal and mechanical properties of these polyesters. This section describes some of the more common polyesters with either the diol or the diacid altered and those which are easily hydrolyzable.

Polyalkylene terephthalates are polyesters with varying lengths of alkyl diols. PET has two methylene units and is referred to as 2GT. Other polyalkylene terephthalates used for polyester synthesis include polypropylene terephthalate (3GT) and polybutylene terephthalate (4GT). 4GT, also known as polytetramethylene terephthalate (PTMT) (**I**), is synthesized by 1,4-butanediol and terephthalic acid. The PTMT synthesis was covered in the original Whinfield and Dickson's 1941 patent, but was not developed for commercial use until much later. In comparison to PET, the longer aliphatic chain length in the diol gives PTMT lower melting (by ~30°C) and glass transition temperatures (near room temperature). PTMT also crystallizes much more rapidly than PET. Its better compressive recovery characteristics have led to its use with PET as side-by-side bicomponent fibers. PTMT has a slightly lower density (1.32 g/cm^3) and is more resistant to hydrolysis than PET.

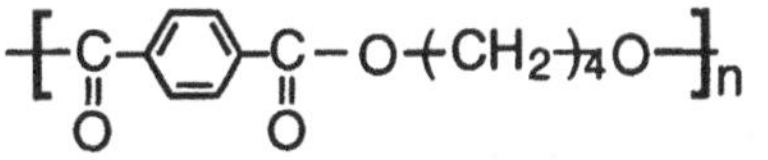

I

Another polyester with a varied diol structure is polycyclohexane terephthalate (**II**), known as Kodel 2 and commercialized by Tennessee Eastman. Structure (**II**) is synthesized from 1,4-cyclohexane dimethanol or 1,4-bishydroxymethyl cyclohexane and terephthalic acid. The cycloaliphatic diol is formed by two consecutive hydrogenation steps from dimethyl terephthalate. The rigidity of the cyclohexane gives this polyester its higher melting and glass transition temperatures by 40°C and 15°C than PET, respectively. Its density (1.22 g/cm^3) and tenacity are lower than PET.

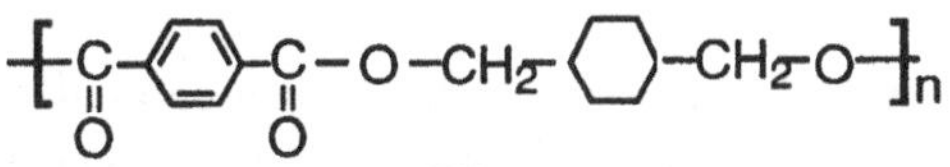

II

Other polyesters contain variations of the diacid structures. Poly(ethylene oxybenzoate) (PEOB) (**III**) with the trade name of A-Tell is polymerized from an intermediate, which is derived from phenol by a three-stage synthesis. The reduced carbonyl–phenol resonance and symmetry lead to lowered melting (224°C) and glass transition (64°C) temperatures and a slightly lowered density (1.34 g/cm^3).

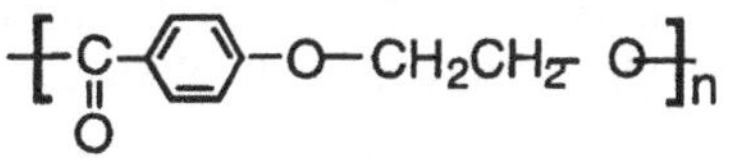

III

Both poly(ethylene diphenoxy ethane-4,4-dicarboxylate) (**IV**) and poly(ethylene naphthalene-2,6-dicarboxylate) (PEN) (**V**) have higher melting (260°C) and glass transition (~110°C) temperatures and much higher moduli than PET, making them suitable for tire cords. Structure (**IV**) was evaluated by ICI and Asahi Kasei and the detailed technical information can be found in Japanese literature. Polypivalolacton (**VI**) was developed by Shell and Kanegafuci. Du Pont and ICI were also active in its development. Its fast crystallization behavior makes the melting processing hard to control.

$$\left[\!\!\begin{array}{c}\!\!\underset{O}{\overset{\parallel}{C}}\!\!-\!\!\bigcirc\!\!-O\!-CH_2CH_2\!-\ O\!-\!\bigcirc\!\!-\underset{O}{\overset{\parallel}{C}}\!\!-O\!-CH_2CH_2\!-\ O\!\!\end{array}\!\!\right]_n$$

IV

$$\left[\!\!\begin{array}{c}\!\!\underset{O}{\overset{\parallel}{C}}\!\!-\!\!\bigcirc\!\!\bigcirc\!\!-\underset{O}{\overset{\parallel}{C}}\!\!-O\!-CH_2CH_2\!-\ O\!\!\end{array}\!\!\right]_n$$

V

$$\left[\!\!\begin{array}{c}\!\!\underset{O}{\overset{\parallel}{C}}\!\!-\!\!\underset{CH_3}{\overset{CH_3}{\underset{|}{\overset{|}{C}}}}\!\!-CH_2\!-O\!\!\end{array}\!\!\right]_n$$

VI

Poly(L-lactide) **VII** is a hydrolyzable polyester which can be used as absorbable surgical sutures. Poly(hydroxy acetic acid) **VIII** or polyoxyacetyl is another hydrolyzable polyester which hydrolyzes faster than poly(L-lactide). It was developed by American Cyanamid and commercialized through its Davis and Geck subsidiary under the trade name Dexon. This polyester is produced by cyclic dimerization of hydroxy acetic acid called glycollide, followed by cationic ring-opening polymerization of the cyclic dimer. This polyester is known as polyglycolide in the United States and polyglycollide in the UK. The Ethicon sutures produced by Johnson and Johnson are based on poly(hydroxy acetic acid) copolymerized with 10% of poly(L-lactide) (Vicryl). Johnson and Johnson also developed polydioxanon or poly(2-oxyethoxy acetate) (PDS) (**IX**), said to be superior to Dexon due to its lower modulus. Poly(hydroxy butyric acid) (PHBA) (**X**), developed by ICI, is made by bacteria grown in a medium containing glucose and methanol.

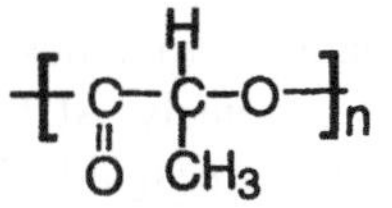

$$\left[\begin{array}{c} \text{H} \\ | \\ \text{C}-\text{C}-\text{O} \\ || \quad | \\ \text{O} \quad \text{CH}_3 \end{array} \right]_n$$

VII

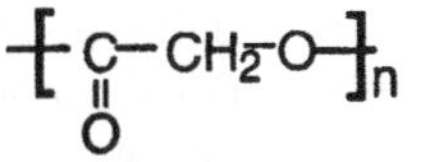

$$\left[\begin{array}{c} \text{C}-\text{CH}_2-\text{O} \\ || \\ \text{O} \end{array} \right]_n$$

VIII

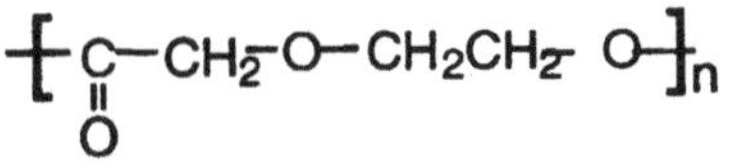

$$\left[\begin{array}{c} \text{C}-\text{CH}_2-\text{O}-\text{CH}_2\text{CH}_2-\text{O} \\ || \\ \text{O} \end{array} \right]_n$$

IX

$$\left[\begin{array}{c} \text{H} \\ | \\ \text{C}-\text{CH}_2-\text{C}-\text{O} \\ || \quad\quad | \\ \text{O} \quad\quad \text{CH}_3 \end{array} \right]_n$$

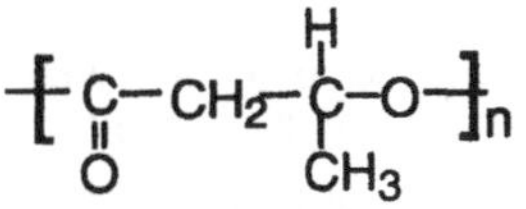

X

C. Copolyesters

In addition to changing either the diol and/or the diacid component in the PET structure, copolymerization is another approach to alter the chemical compositions of polyesters, thus their physical properties. Copolyesters have been developed to serve different purposes. The addition of a comonomer alters the crystallization behavior and thus the microstructure and macrostructure of the fibers. Copolymerization with monomers containing targeted chemical compositions produces copolyesters with added chemical functionality. For instance, receptive sites for dyes can be designed as part of the comonomer structure. Other comonomer structures may improve elastic properties. Random copolyesters can be synthesized by substituting either the acid or glycol constituent of the monomers. Block copolymers are less easily controlled because the ester interchange reaction in the melt tends to drive the block arrangement to a random structure. The copolyester can be synthesized by ester exchange of the dimethyl ester with a more than 2 mol equivalent of the glycol to form trimers, followed by polycondensation eliminating the glycol.

The content of the comonomer is usually limited to 8–10 mol% to avoid detrimental effects on the fiber mechanical properties. In cases, where thermal or

adhesive properties are primary concerns and mechanical properties are secondary, copolymer compositions can be modified more extremely. At 10–30 mol%
of copolymerization, the thermal and physical properties are significantly altered.
Melting temperatures are lowered with increasing levels of the comonomers, but
irrespective to the structure of the substituents in the comonomers. Glass transition temperatures, on the other hand, depend on both the comonomer levels and
the substituent structures. At above 30–35 mol%, the melting points of these
copolyesters become broader and less distinct. Highly modified copolymers can
also be used as bicomponent fibers, in either side-by-side or, more commonly,
sheath–core configuration.

The most common commercial copolyester fibers are the sulfonated PET
(SPET) fibers which contain 1–3% of 5-sulfo-dimethylisophthalic acid (DMS),
a sodium-substituted diacid, as a comonomer:

$$\left[\!\!\begin{array}{c} C\\ \parallel\\ O \end{array}\!\!\left\langle\!\!\raisebox{1ex}{X}\!\!\right\rangle\!\!\begin{array}{c} C\\ \parallel\\ O \end{array}\!\!-O-CH_2CH_2-O\right]_n \quad , X=SO_3Na^-$$

Although the polymerization reactions are similar, the molecular weight of the
SPET is about one-half [4] to two-thirds [5] of that of PET. The SPET fibers are
of lower strength, but the anionic nature of the comonomer makes the SPET
fibers dyeable with basic and cationic dyes. At the stages of as-spun and undrawn
fibers, the voluminous sulfo-groups make the SPET chains more mobile than
those of PET [4]. Upon drawing, smaller crystals and lower orientation are
formed in the SPET fibers than the homopolymer counterpart. The polar sulfo-
groups in SPET fibers also improve their static properties [6,7]. When finishing
the sulfonated PET containing more than 10% of DMS with $ZnSO_4$, the moisture
contents rise to 3–8% [8]. Other acid substitutes include isophthalic acid and
glutaric acid [9]. The glycol substitutes can be DEG or 1,4-cyclohexanedimetha-
nol. More details on the comonomer compositional effects and fiber structure
and properties can be found in a review on ionomeric polyester fibers [5].

III. SURFACE STRUCTURE AND CHEMISTRY

The surface structure (chemistry or morphology) of PET fibers is not as well
understood as the microstructure and macrostructure of the PET fibers as a whole.
The accepted model for melt-spun semicrystalline PET is a microfibrillar structure containing crystalline and noncrystalline regions with many intrafibrillar
molecules and some tie chains connecting the crystalline regions. During spinning and the subsequent processes, some of the chains or chain segments are
preferentially oriented and aggregated by close packing with neighboring molecules into cohesive structural units which are usually referred to as crystallites.

It is generally agreed that polymer chain folding is involved in the development of crystalline domains. The remaining chains exits in a range of packing order. The crystalline regions are composed of closely aligned stacks of chains folding back on themselves, forming the crystals, and the noncrystalline regions contain the chain folds, chain ends, and tie molecules. This model presents a fiber structure with a high degree of anisotropy or orientation in both the crystalline regions and the amorphous areas. The microstructure model has been supported by experimental data on the mechanical and thermal properties.

It is not known, however, if fiber surfaces can be described by the same microstructure model. This is due partly to the limitations of the direct and analytical methods for surface characterization. The understanding of surface structure require much consideration of the energetic requirements for forming surfaces (i.e., the minimization of surface energy). One of the questions is whether the surface composition of PET fibers resembles the bulk. Assuming minimum end-group effect due to its high molecular weight, the carbon-to-oxygen (C/O) ratio in the repeating unit of the PET structure should be 1.875. If all hydroxyl end groups are located at the surface, the C/O ratio should remain similar. If the surfaces are covered with only carboxylic end groups, the C/O ratio would be 1.5. Based on the relative intensities of C_{1s} and O_{1s} peaks from the x-ray photometric spectroscopy (XPS) or elemental spectroscopy for chemical analysis (ESCA), the C/O ratio of polyester fibers was reported to be 2.92 [10]. This C/O ratio is much higher than the 1.875 C/O of the bulk. This measured C/O ratio indicates that only 64% of the oxygen along the PET chain, or an average 2.6 out of the 4 oxygen in the repeating unit, is located on the fiber surfaces. This observation suggests that not only the polar end groups (—OH and —COOH) but also ester linkages along the chain tend to orient themselves toward the bulk, away from the fiber surfaces. As polar groups contribute toward higher surface energies, a more carbon-rich surface is consistent with energy minimization of the fiber surfaces. In order for oxygen atoms along the PET chains to be imbedded, either rotation and/or folding of the glycol portion has to occur. The lower surface oxygen content of the PET fibers suggests that surface chain segments are largely in the noncrystalline phase.

Although relatively few, both hydroxyl chain ends and carboxylic groups are present in the PET structure, as stated earlier. In fact, the carboxylic end groups of PET produce acidity in the range of 2×10^{-2} to 4×10^{-2} mol/kg [11]. When compared with the aliphatic hydrocarbon structures in the olefins, these polar end groups and the ester groups along the PET chains contribute to the somewhat more polar nature of PET. For instance, the water contact angles of polyethylene and polypropylene, are about 95° [16]. The water contact angles of PET, fibers of films, are in the range of 70°–75°. Poly(hexamethylene amide) or Nylon 66 fibers, due to the more polar amide linkage along the chain, have a lower water contact angle of 61.4°.

Titration is common for the determination of the surface polar groups. However, the amounts of hydroxyl and carboxyl end groups on the PET fiber surfaces are extremely low due to the high molecular weight, making common titration methods not as useful. Two derivation methods have been reported for the determination of the surface hydroxyl chain ends, on untreated and base-hydrolyzed PET films [13]. One method involved activation of —OH with p-toluenesulfony chloride (Cl-Ts) in pyridine. The tosylated group was then displaced with ^{3}H-labeled lysine in aqueous solution. Liquid scintillation counting (LSC) of the radioactivity gave derived hydroxyl functions on the surface. The other method coupled a diisocyanate spacer on the hydroxyl group of PET, then reacted it with ^{3}H-labeled lysine followed by LSC. These LSC assay results were confirmed by the XPS analyses of these derivatized PET surfaces. With a higher surface-to-volume ratio of fibrous materials, these methods are potentially useful for improved quantification of hydroxyl chain ends on PET fibers.

IV. SURFACE ENERGY AND WETTING CONTACT ANGLE

Surface free energy of a solid is defined as the energy required to create a new surface. Theoretical surface free energy of a polymer can be calculated by interatomic forces. However, actual surface free energy of a polymer may not be consistent with the theoretically derived one because of the variations in processing and heat treatments. Furthermore, polymers are dynamic and the behavior of chains on the surface may vary depending on the surrounding environment.

Surface energetics of solids determine the surface and interfacial phenomena, including chemical reactivity and diffusion, detergency (adsorption, desorption), wet processing (dyeing and finishing), and adhesion. Wetting behavior of solids in liquids has been closely related to the surface free energies of the solids. Surface energies of solids can be determined by wetting contact angle measurements of solids in liquids using sessile drop, horizontal meniscus, and Wilhelmy techniques [14]. On fibrous solids, the most well-known method is the Welhelmy technique [15]. In this method, the cosine wetting contact angle can be calculated from the measured wetting force value (F) according to

$$F = P \gamma \cos \theta \tag{1}$$

where P is the fiber–liquid interfacial dimension and γ is the liquid surface tension. A low contact angle θ indicates the solid is more wettable by the liquid and a high contact angle indicates the opposite.

An experimental method has been described for deriving the wetting contact angle of the fibers from fabric and yarn measurements [12,16]. In this method, the wetting and wicking components of a steady-state measurement of a fabric are decoupled, allowing the derivation of the liquid wetting contact angle and retention value. This experimental approach and decoupling method has been tested on fabrics with different fiber contents. The wetting contact angles derived from fabric measurements have been confirmed to be the same as those from the single fiber measurements. Due to the easier handling of fabrics than single fibers and the larger force-scale requirements, the fabric measurements are much more versatile than the single-fiber measurements. This experimental approach also has the advantage of providing information on the liquid transport in the fabrics and is applicable for obtaining the liquid wetting contact angle and absorbency of porous materials with different geometric configurations [17]. With this method, the water contact angles of PET, SPET, and microdenier polyester fabrics in water are reported to be 75.8°, 63.9°, and 75.5°, respectively [18]. These data show that the regular denier and microdenier PET fibers have the same water wettability and are less hydrophilic than Nylon 66 fibers ($\theta = 61.4°$) [16] but more hydrophilic than polyethylene fibers ($\theta = 95.6°$) [19]. The anionic nature of comonomer in SPET, even at a very low level, makes these fibers more hydrophilic.

Surface energies of a solid are calculated from the equilibrium contact angle between a solid and a liquid. Two questions arise concerning the accuracy of the equilibrium contact angle measurements; one relates to the dependency of the dynamic contact angles on the moving velocity of the solid–liquid–air boundary, or the liquid movement; and the other relates to hysteresis. It has been shown that by varying the liquid velocities, a velocity range can be obtained where the wetting force is independent of the velocity, and thus an accurate derivation of the equilibrium or thermodynamically significant Young's contact angle. The velocity range where the wetting force was independent of velocity has been reported to be 0.2–0.4 mm/min. for smooth glass slides [20] and nylon fibers [21]. Contact-angle hysteresis is the difference in the contact angles between the advancing (during immersion) and receding (during emersion) wetting measurements of a solid in a liquid. Under the circumstances where no chemical interactions (reaction, swelling, or penetration) occur, contact-angle hysteresis can be caused by surface heterogeneity, i.e., chemical (including contamination) or physical (roughness). Wetting force measurements of PET fibers in a water/n-hexane mixture gave an advancing contact angle (θ_a) of 110° and a receding contact angle (θ_r) of 91° [22]. These water/n-hexane contact angles again show that polyester is less hydrophilic than nylon ($\theta_a = 95°$, $\theta_r = 64°$), but more hydrophilic than polyethylene ($\theta_a = 159°$, $\theta_r = 139°$).

Using the modified Young's equation [23] the dispersive γ^d and polar γ^p of the surface free energy of a polyester fiber determined in n-alkane/water systems are 28.3 mJ/m^2 and 11.7 mJ/m^2, in the advancing and receding modes, respectively [21]. The overall surface energy γ of PET fibers is 40.0 mJ/m^2, which is considerably lower than those of nylon (52.9 mJ/m^2) but higher than polyethylene

(23.1 mJ/m^2) fibers [24]. These derived surface energies are consistent with the wetting contact angles in water and water/n-hexane.

V. SURFACE EFFECTS FROM FIBER SPINNING

A. Fiber Size and Cross-Sectional Shape

The size and cross-sectional shape of fibers, normally described by denier or tex, have direct effects on the surface properties of fibers as well as many yarn and fabric properties. The size or diameter of fibers in a fibrous product determines its total surface area and the proportion of its surface, often termed surface-to-volume ratio. The surface-to-volume ratio of a fibrous product is $2/r$ or $2\pi rl/\pi r^2 l$, where r is the fiber radius and l the total fiber length at given mass. In other words, the surface-to-volume ratios increase linearly with reducing fiber diameters. Properties that depend on the proportion of surface areas of fibrous products (i.e., dyeing and chemical reactions) are expected to vary with fiber size. Although the rate of chemical reactions should be enhanced, by higher surface-to-volume ratios in fibers with smaller diameters, the final results may not be equally positive. For instance, more dye is needed to achieve the same shade in dyeing fabrics containing microdenier fibers than those of regular denier fibers. This is because the higher surface-to-volume ratio of smaller fibers enhances dye adsorption and, thus, dye uptake. The increased curvature of fibers with decreasing fiber diameters, on the other hand, increases light reflectance, resulting in a lighter appearance. Higher curvatures in smaller fibers result in higher light reflectance and a lighter appearance. Consequently, larger amounts of dye are needed to dye fabrics containing microdenier fibers than those with regular denier fibers. Furthermore, the much higher refractive indexes and birefringence of PET make achieving dark shades of colors an even greater challenge for products made of microdenier PET fibers.

The shapes of fiber cross sections also affect their appearance and those properties that are associated with fiber packing in the fibrous materials, such as thermal insulation, moisture, and liquid transport properties. Fibers with multilobed cross sections display increased luster, improved handle, and reduced pilling tendency. The shape factor is the ratio of the fiber perimeter of any cross-sectional shape to that of a circular shape at the same denier per filament (dpf) value. Deviation from the circular cross-sectional shape of a fiber gives a shape factor that exceeds one. For instance, square cross-sectionally shaped fibers have a shape factor of 1.13 or 13% higher surface area than circular-shaped fibers of the same denier. Therefore, increasing the shape factors or irregularity of fiber cross-sectional shapes is another way to increase surface-to-volume ratio. Hollow fibers not only give more bulk and cover factor, improve bending resistance and crimp rigidity, and reduce pilling, but they also contain substantial internal surfaces which scatter light and allow for the improvement of other properties. Decreased thermal conductivity has been shown in woven fabrics made of grooved hollow continuous polyester filaments [25]. The improved thermal insolation is due to the higher volumes of air contained in these fibers.

Changing the size and shape of fibers also affects the pore structure, and thus the liquid transport properties of fabrics. The spontaneous fluid transportation is related to the capillary pressure P within the pores:

$$P = \frac{2\pi r \gamma \cos \theta}{\pi r^2} = \frac{2\gamma \cos \theta}{r} \qquad (2)$$

where r is the pore radius, γ is the liquid surface tension, and θ is the liquid–fiber contact angle. For the same liquid and fiber material, the γ and θ values stay constant. Therefore, the smaller the pores, the higher the capillary pressure for transporting liquids. By reducing fiber diameters and/or adding small pores to fiber surfaces, improved liquid wicking can be expected. Polyester fibers with irregular cross-sectional shapes and deep grooved surfaces have been shown to spontaneously transport fluids [26]. These fibers (6–15 denier) have surface areas 2.3–2.8 times those of same denier fibers with round cross sections.

B. Drawing and Heat

Drawing and heat setting are common fiber-formation processes to orient and crystallize the polymer chains along the fiber axis. These processes have been reported to also affect surface energies of PET fibers [27]. Drawing up to 5 draw ratios has been shown to increase the dispersive component of the surface free energy of PET fibers, but does not alter the polar component of the surface energy. Heating under tension also increases the dispersive surface energies of both undrawn and drawn fibers at increasing temperatures up to 140°C. The dispersive surface energies then decrease with further increases in temperatures above 140°C. The polar surface energies remain unchanged with heat except for a slight increase at 190°C, which corresponds to an increase in the $-C-O-$ groups on fiber surfaces determined by ESCA. The authors attributed the increased dispersive surface energies to the increased atomic density at the fiber surface. The authors also speculated that the decreasing dispersive surface energies of fibers treated above 140°C were due to the decreased surface atomic density in spite of the increase in bulk density.

Because there is no direct evidence of the changing surface density, other surface phenomenon may also contribute the changing dispersive surface energies. Further examination of the ESCA data shows that the fibers with a 5 draw ratio have higher $C-C$ (72.2% versus 70.5%) and a lower $C-O$ (15.0% versus 16.7%) peak intensities than the undrawn fibers. It appears that the dispersive component of the PET fiber surface energy may be related to the proportion of the nonpolar segment of the PET chains on the fiber surfaces. For fibers heat treated at 190°C, a higher dispersive surface energy for the undrawn is also accompanied by a higher C—C (65.9%) than that with a 5 drawn ratio (64.2%). The source of fiber drawing and heat effects on the dispersive surface energy of PET is unclear and requires further work.

VI. SURFACE MODIFICATION

Surface treatments after the formation of fibers can generate well-defined and specific features on fiber surfaces. Much research effort has been dedicated to improving the hydrophilicity of PET because hydrophobicity of PET contributes to some of its less desirable properties such as poor wetting and soil-release behavior in aqueous liquids, attraction to oily soils, low adhesion to rubber and plastics, and the tendency for static electricity buildup. Surface energy or polarity of a polymer can be increased by incorporating hydrophilic compounds by means of copolymerization/addition/blending and surface modification by coating, grafting, and reactions.

For efficient wetting and spreading of a liquid on a surface, the surface tension of the solid, γ_{solid}, must be equal or greater than that of the liquid, γ_{liquid}. At 20°C, the surface tension of water (γ_{water}) is 72.8 mJ/m^2 [28]. It is apparent that the surface tension of PET, γ_{PET}, is far below γ_{water}, using either the earlier mentioned surface tension of PET (40.0 mJ/m^2) or the one (44.6 mJ/m^2) from the *Polymer Handbook* [29]. In order for water to wet PET, the γ_{PET} has to be increased to be close to or above γ_{water}. Because adhesive bonding of solids has been positively associated with their surface energies [30], improvement of adhesion usually involves ways to increase surface energy. The most effective way to change the surface energy of polymers is to change their surface chemistry. Often these chemical changes are also coupled with the modification of surface morphology/topography and the elimination of impurities and/or weak boundary layers.

Chemical modification is by far the most industrially feasible way to impose significant changes to the chemical nature of PET. Reactivity rules of organic synthesis, established for reactions performed in the homogeneous phase, are often not applicable to solid–liquid interface reactions such as those employed in surface modification of fibers. This is because surface wet chemistry depends on the solid–liquid interface, which often changes due to altered wetting and adsorption properties of the modified surfaces and varied dissolving properties of the liquid phase surrounding the solid. The following subsections review the recent literature on the surface modification of PET by various means including alkaline and enzyme hydrolysis, plasma, grafting, and excimer UV laser.

A. Alkaline Hydrolysis

Alkaline hydrolysis is one of the most documented methods for modifying the chemical and physical characteristics of polyester fabrics. The nucleophilic attack of a base on the electron-deficient carbonyl carbon in PET causes chain scissions at the ester linkages along the PET chain, producing carboxyl and hydroxyl polar end groups. The increased surface polarity leads to better wettability and soil-release properties. Aqueous alkaline hydrolysis of PET has shown to be topo-chemical. With the progressive alkaline hydrolysis, the PET chains on the fiber surface are etched away and the fiber diameter is reduced, producing fabrics with a softer and more silky hand. Numerous studies have been reported on aqueous alkaline hydrolysis of PET and those up to 1989 can be found in an earlier review [31] and are not repeated here. Those which were not cited in that review and the more recent work are included here.

The mass loss from alkaline hydrolysis has been of interest to many because it indicates the extent of hydrolysis and reduced fiber dimension. Weight loss from aqueous alkaline hydrolysis of PET has been reported to be linearly related to time, with the slope of this relationship dependent on the base concentration [32]. Based on heterogeneous kinetics, however, dissolution of PET in an alkaline medium was tested by a physicochemical model in which the dissolution rate was proportional to the fiber surface area and the hydroxyl ion concentration to a certain power [33]. The derivation of this model shows the weight loss not to be linearly related to time. Dissolution of PET in sodium hydroxide in the $0.5M/$dm^3 to $3.75M/$dm^3 concentration range at 100°C has been experimentally deter-mined to be a first-order thermodynamic process with respect to fiber surface area and hydroxyl ion concentration [33,34]. This physicochemical model predicts a deviation from linearity in the mass loss versus time relationship and assumes a decrease in the dissolution kinetics with time. The dissolution rate constant of this kinetic relationship depends on temperature and satisfies the Arrhenius equa-tion. The activation energy and preexponential collision frequency factor are in-dependent of the OH$^-$ ion contents in the system. The presence of a cationic surfactant has been shown to increase the activation energy and preexponential collision frequency factor [34]. It is evident that surfactants affect the transport of the reactants at the PET–water interface and their energy state.

It has been shown that alkaline hydrolysis of PET can be controlled to alter fabrics at varying levels (i.e., from surface hydrolysis to extensive removal of the constituent polymer) [35]. Surface hydrolysis causes chain scission to form polar hydroxyl and carboxylic groups and increases surface polarity and hydrogen-bonding capacity with water molecules, thus increasing water wettabil-ity. Surface hydrolysis does not impose significant changes on the overall geom-etry or pore structure of the material. As hydrolysis progresses, fiber surfaces are etched away and alteration of fabric pore structure with increased severity of hydrolysis is expected. By varying the extent of hydrolysis, modification of fiber surface wettability alone or improved surface wettability combined with altered fabric pore structure is possible.

The progression of surface modification to eventual alteration of pore structure of polyester fabrics by aqueous sodium hydroxide ($3N$) has been investigated on regular and microdenier PET fabrics [35]. The mass loss, porosity, and thickness reduction on these fabrics increase with increasing hydrolysis temperature or time. The optimal hydrolysis condition to achieve the highest wettability on both types of fabrics was at a $3N$ NaOH concentration and 55°C. At higher hydrolysis temperatures, further enlargement of the fabric pore structure was observed with no further improvement in wetting. Aqueous hydrolysis improved the wettability of the microdenier PET fabrics to a greater extent than that of the regular PET. The improved surface wetting property plays a primary role in improving water absorbency of these fabrics. With the considerably varied wetting properties and pore structures among these PET fabrics, their water absorbancy increased linearly with improved water wettability or increasing cosine water contact angles. The main deviation from this relationship has to do with the quality of the pore structure. The pore-size distribution and pore connectivities are also crucial to the improved water-retention properties. In the cases where the distribution of pores is not optimal or if the pores are not well connected, the improved wetting properties of the fibers cannot be fully utilized in transporting and/or liquid retention.

B. Enzyme Hydrolysis

Enzymes are natural catalysts that aid reactions in living systems. They have long been used to process and/or modify natural products such as food, feed, and pharmaceutical ingredients. Enzymes have also been in use for fiber processing, such as in degumming of silk and desizing of cotton. The recent surge in enzyme use in textile processing reflects the interest and need for discovering more environmentally sustainable chemical finishing alternatives as well as the significant advancements in biotechnology. Most of the recent work on enzyme processing of fibers has involved mainly cellulosic (wood, cotton, bast-linen, regenerated) and protein (wool and silk) fibers [36].

Hydrolytic enzymes have also been successfully applied in organic synthesis. Esterases and lipases catalyze the hydrolysis and formation of ester and amide bonds, respectively. Both have been used to catalyze synthetically useful reactions, such as ester hydrolysis, esterification of alcohols and acids, transesterification of alcohols and esters, intraesterification of esters and acids, and transfer of acyl groups from esters to other nucleophiles such as amines, thiols, and hyroperoxides. Lipases are known to catalyze the hydrolysis of lipids of fatty acids and glycerol at the lipid–water interface. It is therefore conceivable that hydrolyzing enzymes may also catalyze the hydrolysis of the ester linkage in PET.

Preliminary evidence of enzyme-catalyzed hydrolysis of PET has been demonstrated by the use of lipases to improve the hydrophilicity of several polyester fabrics [18]. These hydrolyzing enzymes improve the water wetting and retention properties of the polyester fabrics with negligible changes in fabric mass and pore structure. For instance, a 10-min reaction with a porcine pancreas lipase (ICN Biochemical, 9001-62-1) (1 g/L, pH 8.0 buffer, 35°C) reduces the water wetting contact angle of PET from 75.8° to 38.4° ($\pm 2.5°$) and increases the water retention from 0.22 μL to 1.06 μL per milligram of fabric. This is superior to the optimal water contact angle of 65.0° ($\pm 8.0°$) and water retention value of 0.32 (± 0.01) μL/mg produced by aqueous alkaline hydrolysis of the same PET fabric under the optimal condition ($3N$ NaOH at 55°C for 2 h) [35]. In fact, the enzyme-hydrolyzed PET fabric has equivalent wettability to that of scoured cotton fabric. Enzyme hydrolysis significantly increases the water absorbency of the PET fabrics to reach 74% of the pore capacity in the fabric.

Optimization of the enzyme reaction conditions shows that the enzymatic hydrolysis can be effective under much more moderate conditions than alkaline hydrolysis [i.e., shorter reaction time, ambient temperature (25°C), and under near neutral conditions without the use of a buffer]. The improved water wettability is accompanied by full-strength retention as compared to the significantly reduced strength and mass loss from alkaline hydrolysis. These hydrolyzing enzymes are also effective in improving the wetting and absorbent properties of sulfonated polyester and microdenier polyester fabrics. Under aqueous conditions, these lipase enzymes catalyze hydrolysis of the ester linkage in PET, whereas water acts as a nucleophile to displace the enzyme to form the alcohol and acid end groups. Washing in water and storage in air for up to 4 months do not change the wettability of the hydrolyzed fibers.

C. Low-Temperature Plasma

Plasma glow discharge can be generated from the action of either high-temperature, electric, or magnetic fields. High-temperature plasma causes fusion of polymers and is not feasible for polymer applications. Low-temperature plasma is generated from electrical discharge, rather than high temperature, under low pressure (~ 1 torr or 133 Pa). The free electrons gain energy from the imposed electric field and lose energy through collisions with gas molecules. In low-temperature plasma, an electrically neutral mixture of exited species, such as ions, radicals, electrons, and metastables are formed. These chemically active species along with the accompanied ultraviolet light are energetic enough to rupture chemical bonds and initiate reactions. The low-temperature plasma is much less penetrating than the high-frequency radiations such as γ radiation or x-rays; it therefore affects only the surface layer of a material, but with much greater intensity.

Generally, two types of surface modification can be induced by plasma. The first type includes chain scission, cross-linking, or intermolecular and intramolecular reactions of polymer chains themselves. The second types are plasma-initiated surface reactions (polymerization, grafting, and/or coating) with chemical species in the surroundings. The former is conducted in nonpolymerizable precursors, most commonly noble gases, nitrogen, oxygen, hydrogen, and ammonia. In plasma-initiated reactions, polymerizable and/or reactive compounds can be introduced before, during, or after the plasma exposure. These surface reactions usually lead to the deposition of a thin layer of the reaction products. The major advantages of these plasma processes over the chemical ones are that they affect only the material surfaces and their great intensity requires only a very short time to achieve a significant effect, thus conducive to continuous processes. The plasma effects are proportional to the amount of active species, therefore to the conditions. Generally, plasma effects increase with increasing power and time, and also with decreasing pressure. The low-pressure condition for low-temperature plasma, which used to be a requirement and challenge for large-scale and continuous processes, is no longer a requirement, as discussed later in this section.

It is well recognized that low-temperature plasma is very effective in modifying surfaces of polymers and fibers [37]. The resulting chemical and physical changes lead to changes in surface energies (wetting, adhesion, bonding), friction coefficient, permeability, and biocompatibility. On PET, low-temperature plasma has been shown to be highly effective and efficient in improving the surface water wettability [38], moisture content [39], and water uptake of polyester fabrics [40]. These improved properties are generally enhanced with increasing power and time. Under the optimal conditions, water wettability of the plasma-treated PET fabric (Y.-L. Hsieh, unpublished data) is similar to that of scoured cotton fabric [35]. The increased moisture content (e.g., 3.7% at 70% relative humidity) significantly reduces the electrical resistivity of PET from 1.5×10^{13} to 5.0×10^7 Ω [39].

Short exposure of plasma at low power levels incorporate polar groups on the PET fiber surfaces and contribute to improved water wettability with negligible ablation. Increased oxygen content on fiber surfaces is known to occur following plasma treatments in either nitrogen or air atmosphere [38]. Incorporation of nitrogen onto polyester fabric surfaces has been reported via either nitrogen or air plasma [41]. The incorporation may involve either substitution along the PET chain or formation of polymeric deposits. Substitution is more likely to occur at the methylene carbon because of the lower stability than the benzene carbon.

Prolonged exposure and/or at high power levels, however, ablate the surfaces of materials and cause weight loss. The increased crystallinity of plasma-treated PET indicates the ease of ablation of the less ordered or amorphous regions on the surfaces. The ablation and surface effects are also strongly influenced by the type of gaseous environment. The extent of ablation was most severe in air with the observation of higher mass loss and surface roughness on the PET [38]. Less ablation and roughness is produced by nitrogen plasma and even lesser effects produced by argon plasma. In addition to the increased surface roughness, voids are also produced from plasma etching of delustered PET fibers. The voids are usually elongated and oriented along the fiber axis. Three types of void structures (i.e., multibridged, discrete, and indiscreet voids) have been reported [42]. Multibridged voids are the longest with lengths ranging from 0.05 to 0.11 mm. They are found on fibers with higher intrinsic viscosity or molecular weight and appear to be dependent on the volume of trapped air and the draw ratio. The discrete voids are the smallest types that center around the titanium dioxide delusterant particles and are found in fibers with lower amounts of titanium dioxide [38].

It is also evident that the surface effects produced from plasma in nonpolymerizable gases tend to be unstable. For instance, the superior wettability acquired on plasma-treated PET decays with time before stabilizing to levels of considerable improvement [38]. The initial decay of wettability on glow-discharged PET surfaces is attributed to the subsequent reactions with residual free radicals in the environment [43], the reorientation of the hydrophilic groups toward the bulk, and possibly surface contamination.

To maintain low-pressure levels in low-temperature plasma is technically challenging for large-scale and continuous processes, thus disadvantageous. It has been demonstrated that low-temperature plasma can be generated at atmospheric pressure [10,44]. The glow discharge is stabilized by using helium as a diluent and inserting a dielectric layer between the electric plates with a high-frequency energy source. The discharge conditions of atmospheric low-temperature plasma are similar to those of low-temperature glow discharge. Helium/argon and acetone/argon atmospheric low-temperature plasma has been shown to greatly improve the water wettability of polyester fabrics [10]. The ESCA analysis shows a marked increase of O_{1s} from 25.5% on the untreated surface to over 30% on the plasma-treated ones. The improved water wettability was attributed to the incorporation of oxygen on the fabric surfaces in the forms of $-C-O-$ and $-C-O-O-$ structures. The drawback of this process is that helium is costly for industrial applications.

A much larger number of studies have been reported on plasma modification of PET films. Some findings on PET films are discussed here because of their fundamental nature and technical relevance to PET fibers. It has been shown that the nature of the gas present determines whether the plasma treatments increase or decrease the surface energetics of polymers. Surface energies of PET films have been shown to increase when treated in oxygen, nitrogen, ammoniac, argon, and helium, and to decrease in halogenated gases [45]. Plasma etching of polymers is exothermic and is characterized by the thermal activation energy. The thermal activation energy of oxygen plasma on PET film surface has been experimentally determined to be 3.7 ± 0.2 eV per oxygen atom under the steady-state assumption [46]. Treatments of PET films in organic solvents with solubility parameters close to those of PET improved the stability of wettability generated by the plasma [47,48]. This improved wettability was thought to be due to solvent-induced crystallization on the PET surface. Solvent treatments followed by subsequent plasma exposure have shown to have synergistic effects and to reduce the wettability decay of plasma effects. Surface crystallization is believed to reduce chain mobility and restrict the reorientation of the polar groups following the plasma treatment, thus reducing wettability decay.

D. Surface Grafting

Surface grafting of PET can be made possible by radical initiated reactions. Free radicals can be generated on the PET surface by exposure to chemical initiators [49,50] or energy sources, such as UV light [51,52] and plasma [40,43,51,53–55]. In most cases, the intent was to introduce polar groups, thus improving surface hydrophilicity. However, a few also involved surface fluorinatation to enhance hydrophobicity [56].

Polyester fabrics pretreated with polyethylene glycol and then treated with plasma have improved hydrophilicity, thus antistatic and oil-release properties [53]. The low-temperature plasma glow-discharged PET fabrics grafted with acrylic acid (AA) and n-vinyl-2-pyrrolidinone (VP) have been shown to improve moisture regains of PET fabrics up to 3.7% and 8.1% respectively [54]. The water wetting contact angles of these grafted surfaces are around the 40° range and do not change with time (i.e., no wettability decay) [43]. Near-UV induced grafting of PET with acrylic acid (AAc), acryamide (AAm), sodium salt of 4-styrenesulfonic acid (NaSS), and N,N-dimethyl-aminoethyl methacrylate (DMAEMA) has improved surface wettability [52]. Considerable hysteresis was observed on these grafted surfaces with the exception of NaSS-grafted PET which contains bulky substituents. The γ^p of the acrylic-acid-grafted PET film surfaces increased with the extent of COOH groups on the surface, whereas the γ^d was unchanged [49].

Ultraviolet-initiated grafting of hydrophilic vinyl monomers [acryamide, poly(ethylene glycol) methacrylate, 2-acrylamide-2-methyl propane sulfonic acid, dimethyl aminoethyl methacrylate] induced surface hydrophilicity and significantly improved the antistatic properties of PET fabrics [52]. Fibers spun from PET blended with hydrophilic copoly(amide-ether) at a 50:50 ratio, quaternized, and cross-linked have improved antistatic properties [57].

Benzoyl-peroxide-initiated grafting PET fibers with perfluorooctyl-2 ethanol acrylic monomer $C_8H_{17}-C_2H_4-O-CO-CH=CH_2$ increase their hydrophobicity [50]. The fluorinated surface is smoother than the original PET surface, but its wetting hysteresis is higher, probably due to the chemical heterogeneity of the fluorinated grafts. CF_4 plasma has shown to produce excellent water repellence and improve dyeability of PET [40]. Water-repellent properties of cotton–polyester fabrics have also been enhanced through the use of glow discharge with perfluorinated monomers [55].

E. Excimer UV Laser

Excimer UV-laser-induced etching is an ablative photodecomposition (APD) process which has been applied to polymers to improve their wettability, bondability, adhesion, and printability [58]. Most of the excimer laser frequencies used are in the 193-nm to 351-nm wavelength range. One of the main surface features of excimer laser-treated polymers is the existence of a threshold fluence dependent on the UV wavelength which correlates to the absorption coefficient of the polymers. Surface roughness can be controlled by varying the intensity and time of the photoetching process [59]. The principles of laser–polymer interactions as well as irradiation effects with excimer laser on the technical properties of fibers (e.g., particle and coating adhesion or wetting properties) can be found in an earlier review [60].

Excimer-laser irradiation causes chain scission of PET with most of the fragments ejected from the surfaces as gaseous products, leaving only very small amounts of fragments on the surface. Etching rate is nonlinearly related to the fluences below the APD threshold, but linearly related above the APD threshold. The chain-length reduction occurs in the amorphous surface layer. On amorphous PET, a smooth surface is produced from excimer-laser irradiation. Excimer-laser irradiation of biaxially stretched PET usually resulted in a wavy surface structure [61]. Atomic force microscopic examination of laser-irradiated PET surfaces has shown two-dimensional dendrites whose sizes depend on the latent times for nucleation [62]. Surface-enhanced Raman scattering (SERS) of excimer laser (248 nm and 308 nm)-irradiated PET at fluences well below the APD indicates a reduction in the oxygen content or O/C ratio on the surface and decreased polarity [63]. The orientation of the phenyl rings and of the carbonyl groups on the surface are thought to result from melting and amorphization of PET.

Irradiation of polyester fabrics with an excimer laser at a wavelength of 248 nm has shown to improve hydrophobicity, increase dye uptake, reduce dyeing time [46], and reduced fabric glossiness [64]. Water-repellent PET fabrics have been produced in Japan using this technology [65]. The excimer laser (248 nm) has shown to improve the adhesion of 250-denier PET yarns to rubber [66]. The improved adhesion is believed to be caused by the enhanced affinity of the resin to the roughened fiber surfaces.

F. Other Methods

High-energy implantation of PET films (50 μm thick) with oxygen and metal ions can bring about a significant increase in surface hardness measured [67]. FTIR of the uppermost layer of the oxygen-implanted PET shows lowered ratio of the trans (1470 cm^{-1}) to gauche (1450 cm^{-1}) bands, indicating lowered surface crystallinity. Reduction of the 900-cm^{-1} and 1500-cm^{-1} bands also suggests changes in the aromatic ring structure. On the gold-implanted PET, the implanted gold atoms were found at a depth of 0.82 μm. The increased hardness, however, was found to extend to the depths of a few micrometers. The increased hardness on the surface is attributed to the formation of a carbon-rich layer, whereas no clear explanation has been given to the cause for increased hardness beneath the surface.

Gas-phase phosphonylation has also been shown to activate the PET structure [68]. This concept has been applied to produce covalently bonded electrically conductive polypyrrole on PET fabrics [69]. The synthesis involves three steps (i.e., phosphonylation, grafting, and polymerization). The electrophilic substitution of phosphorus oxychloride groups on the aromatic ring of PET was accomplished by exposing PET fabrics to phosphorus trichloride in oxygen. The chlorine atoms in the phosphorus oxychloride then react with 1-(3-hydroxypropyl) pyrrole vapor to give pyrrole-grafted PET via ester linkage. Further exposure to pyrrole vapor in the presense of a ferric catalyst leads to polymerization of pyrrole, forming a layer of conductive polymer.

VII. SUMMARY

Although the structure and properties of polyester fibers have been extensively studied and their relationships to fiber processing are relatively well understood, much less is known about their surface characteristics. This chapter provides a brief review on the current literature of the surface characteristics of polyester fibers. The main focus is on polyethylene terephthalate (PET) which constitutes the majority of the polyester fibers produced and consumed globally. Other fiber-forming polyesters are also mentioned. Certain work on PET films is cited when either the approaches for surface modification or the methods for surface characterization are relevant to PET fibers.

Polyethylene terephthalate fibers have much higher birefringence and refractive index values than most other fibers. These characteristics cause PET fibers to reflect more light and appear lighter. Higher delusterant content and dye uptake are usually needed in PET fibers to achieve color intensities equal to those of other fibers. Any surface modification resulting in roughening the fiber surfaces helps to reduce light reflectance and darken the color.

The surface chemistry, energy, and morphological structure of polyester fibers are among the most essential surface characteristics that determine the functional properties and the aesthetics of the polyester products. The ester links along the PET chains and the much less hydroxyl and carboxylic acid polar end groups contribute to the slightly more polar nature of PET when compared with the highly hydrophobic aliphatic hydrocarbons such as the olefins. PET fibers have measured water contact angles in the 70° to 75° range, compared to the 90° to 98° range for hydrophobic polyethylenes and polypropylene. The overall surface energy γ of PET fibers is 40.0 mJ/m^2, with a dispersive γ^d of 28.3 mJ/m^2 and a polar γ^p of 11.7 mJ/m^2.

Fiber spinning can affect the physical (light, bending), chemical (dyeing and reactivity), and wicking properties of PET fibers by changing the size and shape of fibers. Both drawing (up to 5 DR) and heating (up to 140°C) can also increase the dispersive component γ^d of the surface free energy of PET fibers, but do not affect the polar component of the PET fiber surface energy in general.

The experimentally determined C/O ratio of PET fiber surfaces is higher than the theoretical C/O ratio in PET, indicating that the oxygen atoms in the ester linkages along the chain and the polar end groups ($-$OH and $-$COOH) are not uniformly distributed on the surface of the fibers as in the bulk. This oxygen-deficient surface composition is consistent with minimization of the fiber surface energy and can be attained by either rotation and/or folding of the PET chain. This surface chemical composition suggests that surface chain segments are sufficiently mobile and mostly likely to be noncrystalline.

The hydrophobic nature of PET fibers causes them to display poor water wettability, dyeability, and soil-release properties, low adhesion to rubber and plastics, and tendency for static electricity buildup. Increasing surface energy or polarity of PET fibers is thus logical to improve surface wettability, increase the adhesive bonding, and reduce the build up of static electricity. Because that surface polarity, thus energy, can be most effectively altered by changing surface chemistry, chemistry modification of PET surfaces have been reviewed. These chemical approaches include alkaline hydrolysis, enzyme hydrolysis, low-temperature plasma, surface grafting, and excimer UV laser. Often, chemical changes incurred by these methods are coupled with either the elimination of surface irregularities and weak layers and/or the modification of surface morphology/topography. These recent works have enhanced our understanding on the effects of the various means of modification on the surface structures of PET fibers as well as the chemical compositional effects on their surface energetics. Much in-depth scientific research and technological advancement are still needed for fundamental understanding of the molecular organization on PET fiber surfaces and for developing industrially feasible methods to achieve functional improvement of PET fiber surface properties.

REFERENCES

1. Polyester Textiles, Shirley Conference Publication S51, 1988.
2. J. E. Ford, Polyester Textiles, Shirley Conference, 1988, pp. 1–14.
3. J. E. McIntyre. Polyester Textiles Shirley Conference, 1988, pp. 50–60.
4. D. A. Timm and Y.-L. Hsieh. J. Polym. Sci. B: Polym. Phys. Ed. *31*:1873 (1993).
5. B. R. Rao and K. V. Datye. Text. Chem. Color. *28*(10):9 (1996).

6. VED Chemiefaserwrk Fredrick Eagles. East German Patent No. 126943 (1985); Chem. Abstr. 161,799.
7. Japan Estor Co. Ltd., Japanese Patent No. 60104519 (1985); Chem. Abstr. 197,271 (1985).
8. Ashai Chemical Ind. Co. Ltd., Japanese Patent No. 02026985 (1990), Chem. Abstr. 42406 (1990).
9. J. N. Kerwalla and S. M. Hansen, Polyester Textiles, Shirley Conference, 1988, pp. 77–102.
10. T. Wakida, S. Tokino, S. Niu, J. Kawamura, Y. Sato, M. Lee, H. Uchiyama, and H. Inagaki. Text. Res. J. 67(8):433 (1993).
11. C. A. D'Allo, Ninth Shirley Institute Seminar "Polyester Textiles," 1977.
12. Y.-L. Hsieh. Text. Res. J. 65(5):299 (1995).
13. P. Mougenot, M. Koch, I. duPont, Y. J. Schneider, and J. Marchandbrynaert. J. Colloid Interf. Sci. 177(1):162 (1996).
14. A. W. Neumann and R. J. Good, in Surface and Colloid Science, Vol. 11 (R. J. Good and R. R. Stromberg, eds.), Plenum Press, New York, 1979, pp. 31–91.
15. P. C. Hiemenz, Principles of Colloid and Surface Chemistry, Marcel Dekker, New York, 1986, p. 322.
16. Y.-L. Hsieh, and B. Yu. Text. Res. J. 62(12):677 (1992).
17. Y.-L. Hsieh. Text. Res. J. 64(11):552 (1994).
18. Y.-L. Hsieh and L. Cram. Text. Res. J. 68(5):311 (1998).
19. Y.-L. Hsieh, S. Xu, and M. Hartzell. J. Adhes. Sci. Technol. 5(2):1023 (1991).
20. J. B. Cain, D. W. Francis, R. D. Venter, and A. W. Neumann. J. Colloid Interf. Sci. 94(1):123 (1983).
21. M. Tagawa, K. Gotah, A. Yasukawa, and M. Ikuta. Colloid Polym. Sci., 268:589 (1990).
22. M. Tagawa, K. Yasukawa, M. Gotoh, M. Tagawa, N. Ohmae, and M. Umeno. J. Adhes. Sci. Technol. 6:763 (1992).
23. J. Schultz, K. Tsutsumi, and J. B. Donnet, J. Colloid Interf. Sci. 59(2):272 (1977).
24. M. Tagawa, N. Ohmae, M. Umeno, K. Gotoh, and A. Yasukawa. Colloid Polym. Sci. 267:702 (1989).
25. M. Matsudaira and Y. Kondo. J. Text. Inst. 87(part 1, No. 3):409 (1996).
26. W. A. Haile and B. M. Phillips. TAPPI J. 78(9):139 (1995).
27. Y. Okamura, T. Tagawa, K. Gotoh, M. Sunaga, and T. Tagawa, Colloid Polym. Sci. 274(7):628 (1996).
28. D. R. Lide and H. P. R. Frederikse (eds.), CRC Handbook of Chemistry and Physics, CRC Press, New York, 1996, p. 6–8.
29. J. Brandrup and E. H. Immergut (eds.), Polymer Handbook, Wiley, New York, 1989, p. VI 420.
30. M. Levin, G. A. Ilka, and P. Weiss. J. Polym. Sci. B2:915 (1964).
31. S. H. Zeronian and M. J. Collins. Text. Prog. 20(2):1 (1989).
32. J. Dave, R. Kunar, and H. C. Srivastava. J. Appl. Polym. Sci. 33:455 (1987).
33. N. Kallay, A. M. Grancaric, and M. Tomic. Text. Res. J. 60:663 (1990).
34. A. M. Grancaric and N. Kallay. J. Appl. Polym. Sci. 49(1):175 (1993).
35. Y.-L. Hsieh, A. Miller, and J. A. Thompson. Text. Res. J. 66(1):1 (1996).
36. K.-E. L. Eriksson and A. Cavaco-Paulo (eds.), Enzyme Applications in Fiber Processing, American Chemical Society Symposium Series 687, American Chemical Society, Washington, DC, 1998.

37. M. Ueda and S. Tokino. Rev. Prog. Color. *26*:9 (1996).
38. Y.-L. Hsieh and E. Y. Chen. I&EC Product Res. Dev. *24*(2):246 (1985).
39. N. V. Bhat and Y. N. Benjamin. Text. Res. J. *69*(1):38 (1999).
40. A. M. Sarmadi and Y. A. Kwon. Text. Chem. Color. *25*(12):33 (1993).
41. T. Hirotzu. Text. Res. J. *55*(5):323 (1985).
42. P. R. Blakely and M. O. Alfy. J. Text. Inst. *69*:38 (1978).
43. Y.-L. Hsieh and M. Wu. J. Appl. Polym. Sci. *43*:2067 (1991).
44. T. Yokoyama, M. Kogoma, T. Moriwaki, and S. Okazaki. J. Phys. D: Appl. Phys. *23*:1125 (1990).
45. T. Wakida, H. Kawamura, J. C. Song, T. Goto, and T. Takagishi. Sen-I Gakkaishi *43*(7):382 (1987).
46. R. Ryback, D. Knittel, and E. Schollmeyer. Melliand Textilberichte *73*(12):985 (1992).
47. Y.-L. Hsieh, C. Pugh, and M. S. Ellison. J. Appl. Polym. Sci. *29*:3547 (1984).
48. Y.-L. Hsieh, D. A. Timm, and M. Wu. J. Appl. Polym. Sci. *38*:1719 (1989).
49. T. Kawase, M. Uchita, T. Fujii, and M. Minagawa. Text. Res. J. *61*(3):146 (1991).
50. A. Ghenaim, A. Elachari, M. Louati, and C. Caze. J. Appl. Polym. Sci. *75*(1):10 (2000).
51. F. C. Loh, K. L. Tan, E. T. Kang, Y. Uyama, and Y. Ikada. Polymer *36*(1):21 (1995).
52. E. Uchida, U. Uyama, and Y. Ikada. Text. Res. J. *61*(8):483 (1991).
53. T. Goto, T. Wakita, and I. Tnanaka. Sen-I Gakkaishi *46*(5):192 (1990).
54. Y.-L. Hsieh, M. Shinawatra, and M. D. Castillo. J. Appl. Polym. Sci. *31*:509 (1986).
55. U. Vohrer, M. Muller, and C. Oehr. Surface Coating Technol. *98*:1128 (1998).
56. T. Kawase, T. Fujii, M. Minagawa, H. Sawada, and M. Nakayama. Text. Res. J. *64*:375 (1994).
57. Y. Sano, C. W. Lee, Y. Kimura, and T. Saegusa. Angew. Makromol. Chem. *246*: 109 (1997).
58. S. Lazare, P. D. Hoh, J. M. Baker, and R. Srinivasan. J. Am. Chem. Soc. *106*:4288 (1984).
59. I. Zhang, W. Boyd, and H. Esrom. Surface Interf. Anal. *24*(10):718 (1996).
60. T. Bahners. Opt. Quantum Electron (UK) *27*(12):1337 (1995).
61. A. Brezini. Phys. Status Solidi A: Appl. Res. *135*(2):589 (1993).
62. J. Heitz, E. Arenholz, D. Bauerle, and K. Schilcher. Appl. Surf. Sci. *81*:103 (1994).
63. T. Lippert, F. Simmermann, and A. Wokaun. Appl. Spectrosc. *47*(11):1931 (1993).
64. K. S. Lau, P. W. Chan, K. H. Wong, K. W. Yeung, K. Chan, and W. Z. Gong. J. Mater. Process. Technol. *63*(1–3):524 (1997).
65. H. Tatsuya, *New Fibres*, Ellis Horwood, London, 1993, pp. 94–95.
66. H. Watanabe and T. Takata. Angew. Makro-Mol. Chem, *235*:95 (1996).
67. N. Nishimiya, K. Ueno, M. Noshiro, and M. Satou. Nucl. Instrum. Methods Phys. Res. B *59*:1276 (1991).
68. S. W. Shalaby and M. S. McCaig, U.S. Patent 5,491,198 (1996).
69. D. Tessier, L. H. Dao, Z. Zhang, M. W. Martin, and R. Guidoin. J. Biomater. Sci. Polym. Ed. *11*(1):87 (2000).
70. K. Gotoh, A. Jauskawa, M. Ohkita, H. Obata, and M. Tagawa. Colloid Polym. Sci. *273*:1144 (1995).

3
Frictional Properties of Textile Materials

BHUPENDER S. GUPTA   Department of Textile Engineering, Chemistry, and Science, North Carolina State University, Raleigh, North Carolina

I. INTRODUCTION

Friction is an important property of textile materials that governs the quality and efficiency of processing and the performance of products. Although widely studied, it is still considered to be one of the most complex, least understood, and least controllable of the properties.

Friction arises from interaction between two surfaces. Any factors that affect one or both surfaces affect the force necessary to overcome friction. Numerous studies have been conducted to measure and study the property, either of a textile material against itself or of a textile material against another surface. The factors whose effects have been considered important and studied can be classified into two groups: material related and operation related. Among the former are the fiber morphology (surface features and cross-sectional shape), fiber size, and surface chemical and physical characteristics. The last includes molecular orientation and deformational behavior of the asperities where contacts occur. The operation-related effects include such factors as the normal force pressing one material against another, the size of contact, the speed of sliding, and the environmental conditions. A factor not included in these groups but is used extensively in modifying friction is lubrication.

The field of friction is broad and includes within its scope both the fiber-to-fiber and the fiber-to-other surfaces, in particular metals, behaviors. Covered in this chapter are, however, the studies related to only the first aspect (i.e., the fiber-to-fiber friction). The topic is presented in six major sections. Section II includes a discussion of the concept of friction and a generalized model that can be used to characterize it. The model gave detailed structure to the classical parameter, the coefficient of friction, and the indices a and n of the equation $F = aN^n$. It brought to light the factors, structural as well as procedural, that affected the values of these parameters. Section III includes the experiments conducted and the results obtained on textile fibers and fibrous bundles. The effects studied were those of the structural variables (molecular orientation, setting and annealing, and cross-sectional shape) and procedural variables (testing environment, mode of contact and normal pressure). Augmenting these results are those included in the fourth and the fifth sections, which illustrate the important role friction plays in areas outside the traditional textiles. In the fourth section, the frictional behavior of human hair, the fiber that resembles wool in its chemical and physical structure, and the effects of some of the factors to which the fiber is generally exposed are discussed. The effects studied are those of the chlorine found in swimming pool water and the chemicals contained in the cosmetic compounds used for coloring, bleaching, and permanent waving. In the next section, the role friction plays in governing the behavior of knots tied in surgical sutures is examined. The effects of the suture material, the surface treatments, and the testing conditions and environment are considered. In a study completed recently, the frictional properties of woven fabrics was investigated. It involved frictional measurements on (1) a set of commercially available fabrics, varying broadly in terms of fiber type, weave tightness and construction, and yarn structure and (2) a set of model fabrics, varying systematically in structure. This work is presented in the sixth section. In the seventh and final section, a general summary of the status of the work in the field, including the gaps that still exist in our understanding, and the studies one can undertake to close some of these, are given.

II. CONCEPT AND GENERAL MODEL

Friction is the force that opposes relative motion between two bodies. It is tangential to the surfaces in contact and is related to the force pressing the bodies together. It is largely influenced by the nature and the morphology of the surfaces in contact. Attraction of one surface to another leading to bonding or adhesion over areas of contact, engaging of high and low regions of rough surfaces requiring work to pull one surface over another, and/or deformation of soft regions by the asperities of hard regions are generally considered responsible for the observed phenomena. On planar, relatively smooth surfaces, the main mechanism widely thought to be operating is the formation of junctions due to adhesion at the points of contact; these are ruptured or sheared by application of tangential force in order to initiate sliding. Because junctions can form almost instantaneously, force must be used to maintain sliding. In viscoelastic materials, because the time of contact under pressure can affect the area, time becomes a variable and, therefore, the length of time of contact before initial sliding and the speed of sliding start to play additional roles. This is conceptually not the case if the material is either plastic or elastic. Temperature and humidity conditions influence the results if surface properties or morphologies are modified in any way.

A significant progress toward an understanding of the nature of friction in materials in general was made when Bowden and Tabor [1,2] put forward their concept known as the adhesion-shearing theory of friction. According to it, as mentioned earlier, junctions are formed at points of contact which must be sheared for sliding to take place. Ignoring the ploughing force that may exist in bodies with sharp points, they considered the frictional force as being the product of the true area of contact, A, and the bulk specific shear strength of the junctions, S:

$$F = SA \tag{1}$$

For materials which deform plastically (e.g., ductile metals), the value of A could be shown to be given by the ratio of the normal force, N, to the yield pressure, P_y, of the material. Therefore,

$$F = \left(\frac{S}{P_y}\right) N = \mu N \tag{2}$$

This is the classical equation which supports Amontons' first law and is used to define the parameter the coefficient of friction, μ, given by the ratio of the frictional force, found from experiments, to the normal force, imposed during measurements.

It is well established, however, that Eq. (2) with the implication of μ as a material property is not valid on fibers. In such materials, which behave viscoelastically, μ is found to be a function of both the normal force and the apparent area of contact. Fitting of experimental data to a variety of models has shown that a concise empirical equation that can effectively describe the behavior of such materials is as follows [3]:

$$F = aN^n \tag{3}$$

In this, a and n are constants whose values, determined by fitting of experimental data to the model, characterize the frictional behavior. Several authors [2–11] have conducted experiments to determine the values of these constants and have found that the values vary with the material and the test conditions.

In order to characterize friction in materials in general and give theoretical meanings to these constants, Gupta and El-Mogahzy proposed a structural model [12]. They assumed that the compressional behavior of materials (pressure versus contact area) was given by the following general relation:

$$P = KA^\alpha \tag{4}$$

Behaviors of a variety of types could be represented by varying the values of K and α (Fig. 1), the former regarded as a stiffness or hardness factor and the latter as a shape factor.

According to the adhesion-shearing concept, contacts take place only at the tips of junctions (Fig. 2). The total normal force N is distributed over these points, as W_1, W_2, and so on. Deformation occurs at these points until the pressure on each reduces to the point that it can be supported elastically (Fig. 3). The intersections of the isoload curves of the asperities with the pressure–area curve determine the areas of contact at the points. These must be summed in order to determine the value of A in Eq. (1). For an asperity, i,

$$W_i = P_iA_i = K(A_i)^{\alpha+1}$$

or

$$A_i = K^{-\nu}W_i^{\nu}$$

$$A = K^{-\nu} \sum_{i}^{m} W_i^{\nu} \tag{5}$$

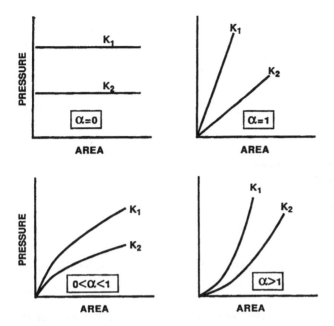

FIG. 1 Pressure–area curves obtained with different values of K and α ($K_1 > K_2$). (From Ref. 12.)

where $v = (\alpha + 1)^{-1}$ and m is the total number of asperities making contact. In order to determine the sum in Eq. (5), it is necessary that we know the way the stress is distributed over the contact region. Considering three different types of distributions (conical, semispherical, and rectangular), the authors showed that the sum can be given by

$$\sum_{i}^{m} W_i^{v} = C_m m^{1-v} N^v$$

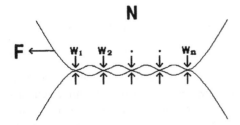

FIG. 2 Distribution of load over points of contact.

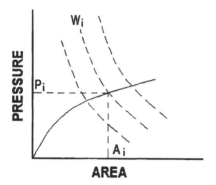

FIG. 3 Determination of the areas of contacts of asperitie.

This, when substituted in Eq. (5), yields the following equation for the area A:

$$A = C_m K^{-v} m^{1-v} N^v \tag{6}$$

In Eq. (6), C_m is a model constant whose value varies by a maximum of 8% among the distributions considered and is 1 if the distribution is rectangular or uniform. (Its value is also 1 if $v = 1$.) This equation is plugged in Eq. (1) to give the following for F:

$$F = S C_m K^{-v} m^{1-v} N^v \tag{7}$$

A comparison of Eq. (7) with Eq. (3) yields the following structures for the empirical constants n and a:

$$n = v \quad [\text{or} = (\alpha + 1)^{-1}] \tag{8}$$

$$a = S C_m K^{-n} m^{1-n} \tag{9}$$

Substitution of Eq. (8) in Eq. (7) gives the following for the frictional force F:

$$F = S C_m K^{-n} m^{1-n} N^n$$

Defining μ as given by Eq. (2), one obtains a model for μ as follows:

$$\mu = S C_m K^{-n} m^{1-n} N^{n-1} \tag{10}$$

The above models bring out the factors that affect the values of the frictional force and the coefficient of friction and of the indices a and n determined empirically. Some of these factors are material related and the others are related to the morphology of contact and the nature of stress distribution. A brief discussion follows which highlights the general effects of some of the important factors.

A. Normal Force

If a material deforms plastically [i.e., $n = 1$ ($\alpha = 0$)], then the normal force N has no effect on μ and Eq. (10) reduces to the dimensionless quantity (S/K). For all other materials, an increase in N causes a decrease in μ.

B. Mechanical Behavior of Junctions

The three factors which govern the mechanical behavior of junctions are the bulk specific shear strength of the junction, S, the hardness factor, K, and the shape of the pressure–area curve, given by the value of the constant n (or α). Although no values have been given for S in the literature, its value can be expected to vary with the chemical nature of the material and with any physical factors, such as molecular weight, molecular orientation, and crystallinity, that modify molecular interaction between surfaces.

The parameters K and n affect the value of the area A and, through it, the values of F and μ. If n is unity, the material behaves plastically, as mentioned earlier, and the pressure–area curve is given by $P = KA^{1-1}$ or K. In this case, μ and a become equivalent and are given by S/K. For all other cases, n is less than 1; this means material behaves viscoelastically, the pressure–area curve is given by $P = KA^{\alpha}$, and K, the hardness factor, has the units of force/(area)$^{1/n}$. In this case, for all values of n, μ (and a) decreases with the increase in K.

C. Number of Asperities in Contact

With the normal force maintained constant, an increase in the number of points of contact, m, causes an increase in the area, A, and, through it, an increase in the values of F and μ. With an increase in m, the value of the index a also increases.

D. Other Factors

Some other factors that affect friction are the mode of contact (line, point, or areal) during tests, the morphology of the surfaces (degree of roughness or smoothness), the testing environment (temperature and relative humidity), and the time of contact (time before sliding and speed of sliding). The mode of contact affects the number of points of contact, m, and through it the values of a, A, F, and μ, mentioned earlier. Tests have shown that line contact leads to higher value of μ than does point contact [13]. The areal contact takes place when two flat surfaces are pressed together. The larger the area, the higher the μ.

Differences in surface morphology can affect friction. A rough surface leading to fewer contacts per unit apparent area, or smaller m, should result in lower friction. Two highly polished surfaces contacting each other should give high friction due to large m and, through it, a large area of contact. It can be expected, however, that a small amount of an impurity getting between the surfaces can break the contacts and drastically reduce the frictional resistance.

The changes in the environmental conditions, such as the relative humidity and temperature, can influence friction by affecting the values of the shear strength of the junctions, S, the hardness of the asperities, K, and the number of asperities in contact, m. If the changes in the environmental conditions affect the viscoelastic properties of the contacting surfaces as well, then the value of n will also be affected, which will impact friction.

Finally, because fibers are viscoelastic materials, their properties change with time as the load is maintained. The values of the parameters, that might be affected, are K, n, and m. Although the effects may occur at all speeds, they are likely to be significant at low speeds. It is expected that the lower the speed of sliding, the greater the contact area and, therefore, the greater the coefficient of friction. This change can be expected to result from a decrease in K or α or an increase in m.

III. EXPERIMENTAL STUDIES OF THE EFFECTS OF STRUCTURAL, MORPHOLOGICAL, AND ENVIRONMENTAL FACTORS

Described in this section are the results from experiments conducted at the North Carolina State University to determine the effects of some of the structural, morphological, and environmental factors on interfiber friction. Whenever feasible, an attempt is made to interpret and understand the effects found in light of the theory presented earlier. Among the structural factors investigated are the type of fiber and the orientation of molecules. The main morphological factors studied are the fiber size, fiber shape, and the morphology of contact. The latter is governed by the procedure used in conducting tests (e.g., aligning the fibers such that they produce either the line contact or the point contact). Other procedure-related factors considered are the normal force and the speed of testing. In one of the studies reported, the effect of changing the environment from dry to wet on friction is examined. Temperature is another that has been used as a factor by some workers [14,15].

In developing test methods, it is recognized that fibers make a variety of contacts with each other during processing and end use, but they rub against each other mostly either along their lengths, such as found during drafting, or at right angles, such as it occurs during insertion of a pick in weaving or during stretching of a fabric. Accordingly, the test methods have been developed to test interfiber friction by maintaining either a line contact in which fibers slide along their lengths while pressed together or a point contact in which one fiber slides against another at right angles.

A. Methods

1. Line Contact

The method used is that proposed by Lindberg and Gralen [3]. In this, two fibers or yarns are twisted together a certain number of times; tension on one end of each is maintained constant and that on the other is increased until relative motion sets in. The equation that allows the calculation of the coefficient of friction is:

$$\frac{T}{T_0} = e^{\pi\tau\beta\mu} \tag{11}$$

In this equation, T_0 is the initial tension, τ is the amount of twist inserted, β is the twist angle or the angle at which the axes of the two fibers are inclined with respect to each other, and T is the tension needed to cause slippage. Implicit in this equation is the assumption that μ is a material property, independent of the normal force, N, or the tension, T_0. An equation can, however, be derived for the general case of viscoelastic materials given by Eq. (3) [13]. The modified equation is

$$T = T_0 \exp^{[\pi\tau\beta a(T_0\beta^2/4r)^{n-1}]} \tag{12}$$

where r is the specimen radius. Comparing Eq. (12) with Eq. (11), one obtains the following expression for μ:

$$\mu = a\left(\frac{T_0\beta^2}{4r}\right)^{n-1} \tag{13}$$

Using this model, one can conduct experiments at different values of T_0, determine the corresponding values of μ with Eq. (11), fit the results obtained to Eq. (13), and estimate the values of the indices a and n. For this, it is necessary that the value of the radius r of the test specimen is known (i.e., either measured directly or estimated with an appropriate model).

An apparatus developed by Gupta [16,17] adapted the twist method for use on an Instron tester. The device consists of a metal plate with six frictionless pulleys (Fig. 4) and is held in the lower jaws of Instron. The test specimens are clamped into the upper jaws, passed around the upper four pulleys, twisted by the required number of times, passed over the lower two pulleys, and then tensioned by hanging known weights. When Instron is started, the ends clamped in the load cell are pulled against the opposing forces of friction and the pretensioning weights. The tension builds in the contacting region until the fibers slip. The fibers hold again, then slip, producing a stick–slip profile. From the latter, the values of the tension corresponding to the static and the kinetic frictional forces are read and substituted in Eq. (11) to obtain the corresponding values of the static and the kinetic coefficients of friction, μ_s and μ_k, respectively. The

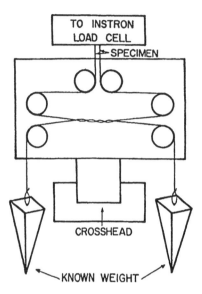

FIG. 4 Schematic of the twist method friction testing device. (From Ref. 16.)

difference $(\mu_s - \mu_k)$ is used as an estimate of the magnitude of stick–slip. By conducting tests at different values of the initial tension T_0 and fitting the data to the model given by Eq. (13), the values of the constants a and n are determined.

2. Point Contact

For this, the most convenient setup corresponds to the Capstan method. In this, a fiber is held fixed in a U-frame and acts as a rigid rod. One end of the second fiber is clamped in the load cell of Instron and the other led over the first and tensioned with a dead weight (Fig. 5) [18]. The angle of wrap and the diameters of the fibers being small, the contact area is small and approximates a point. The U-frame held in the crosshead is traversed away from the load cell generating a friction profile. The behavior is characterized by the well-known Capstan equation:

$$T = T_0 e^{\mu\theta} \tag{14}$$

As done for the line contact, the equation is modified to apply to materials following the behavior given by Eq. (3) [19]:

$$T = T_0\exp^{[\theta a(T_0/r)^{n-1}]} \tag{15}$$

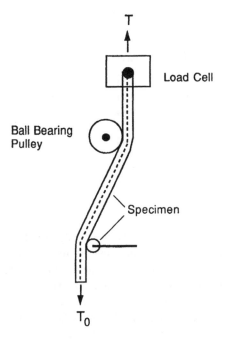

FIG. 5 Schematic of the point-contact device. (From Ref. 18.)

A comparison of Eq. (15) with Eq. (14) yields the following for μ:

$$\mu = a\left(\frac{T_0}{r}\right)^{n-1} \tag{16}$$

Tests are conducted in a fashion similar to those used for the line-contact method and the values of the parameters μ, a, and n are determined. The two devices discussed have provisions for conducting tests with the contact regions submerged in a fluid.

B. Results

1. Effect of Fiber Cross-Sectional Shape

Friction was measured on monofilaments of polypropylene of 18 denier and three different cross-sectional shapes: circular, triangular, and trilobal. The results showed that in both the line- and the point-contact methods, the circular fiber gave a significantly higher value of μ than did the triangular or the trilobal materials, the average difference being about 31% [13]. The values of a and n for the circular fibers were also higher: 28% and 3.6%, respectively, higher than the

TABLE 1   Values of μ, a, and n Obtained in Dry and Wet
Tests Using Line-Contact Method

Fiber	Parameter	Dry	Wet
Polypropylene	$\mu(T_0/r = 328\ g_f/mm)$	0.29	0.34
	a	0.77	0.94
	n	0.83	0.84
Acrylic	$\mu(T_0/r = 87\ g_f/mm)$	0.19	0.23
	a	0.32	0.44
	n	0.68	0.67

Source: Ref. 13.

corresponding average values for the noncircular materials. It was suggested by
the authors that a difference of about 10% in the value of a and 5% in the value
of n could be considered statistically significant. Accordingly, the differences in
the values of μ and a between the two materials were significant. An explanation
given was that the molecules in the lobes of the noncircular, as compared to those
in the circular, were more highly oriented and packed. This led to contact regions
which were stiffer and less deformable in the former than in the latter.

2.   Effect of Testing Conditions

The test devices described earlier allowed the measurements of friction to be
made with the contact region submerged in a fluid. The materials used were
acrylic yarns (unannealed, 200 denier, stretch ratio 2.0) and polypropylene yarns
(1390 denier). Tests were conducted using line contact as the method and water
as the fluid. The results obtained are summarized in Table 1. Although changing
the environment from dry to wet did not affect the value of n, it caused significant
increases in the values of μ and a. Because these fibers were not hydrophilic,
which could attract moisture in their internal structure and swell, a possible explanation given was that water acted as an antilubricant. It removed the finish present
and increased the area of contact between the materials.

3.   Effect of Molecular Orientation

The effect of molecular orientation on friction was measured in multifilament
polypropylene and acrylic yarns which had been given different stretch ratios
[13,20]. In the former, the stretch caused the denier and, therefore, the diameter,
to decrease. In the latter, on the other hand, the jet stretch (which does not significantly affect orientation) and cascade stretch (which affects molecular orientation) were varied such that orientation changed but the final denier remained
constant. The amounts of stretch given to acrylic yarns during processing, the
values of the orientation factor, assessed by a sonic modulus procedure, and the

TABLE 2 Specifications and Values of Sonic Modulus Orientation Factor (f) and Coefficient of Friction μ for Acrylic Yarns Used in the Study

| | Stretch | | | | μ | |
Sample	Jet	Cascade	Total	f	Point	Line
1	2.50	2	5.00	0.69	0.135	0.186
2	1.68	3	5.03	0.73	0.136	0.221
3	1.25	4	5.02	0.76	0.134	0.230
4	1.00	5	5.02	0.77	0.138	0.235
5	0.84	6	5.04	0.78	0.138	0.238
6	0.72	7	5.02	0.79	0.141	0.243

Source: Ref. 13.

values of the coefficients of friction determined by the line- and the point-contact methods, are shown in Table 2. The values of μ for the polypropylene yarn and those of a and n for the polypropylene and the acrylic yarns, are given in Tables 3 and 4, respectively.

The results showed that μ increased with orientation. The values of a and n did not change much in the point-contact method, but the value of a decreased and that of n increased significantly with increase in orientation in the line-contact method. Scanning electron microscopic (SEM) analyses of the surfaces of acrylic yarns showed that the morphology changed gradually with an increase in cascade stretch: The macrofibrils stretched out into alignment, packed more compactly, and the surface became increasingly smoother [20]. This could be expected to lead to more intimate contact (higher m) and thus to larger μ with larger stretch. The increase in n with orientation indicates that the material became more plastic in nature. This could be expected from a possible decrease in the compressive strength of the junctions with orientation. The decrease in a noted was considered

TABLE 3 Coefficient of Friction of Polypropylene Yarns at Draw Ratios (DR) of 1X and 2X Using Line-Contact Method

| | μ | |
T_0/r (g$_f$/mm)	DR: 1X	DR: 2X
100	0.32	0.38
200	0.31	0.37
300	0.30	0.36

TABLE 4 Values of Parameters a and n for Selected Stretch Ratios

Fiber	Stretch ratio	a $(g_f/mm)^{1-n}$		n	
		Point	Line	Point	Line
Polypropylene	1X	0.49	0.77	0.86	0.83
	2X	0.50	0.59	0.88	0.85
Acrylic	2X	0.30	0.32	0.86	0.68
	7X	0.26	0.22	0.88	0.89

Source: Ref. 13.

to be a consequence of the manner in which the parameters μ, a, and n [Eq. (13)] were connected. For given values of T_0/r and β, the net effect of the changes that occurred in the values of n and μ with an increase in orientation was a decrease in a.

4. Effect of Annealing

In one of the experiments conducted, the frictional behaviors of two sets of acrylic yarns, one annealed and the other unannealed, were compared. In preparing for this experiment, unannealed acrylic yarns with properties given in Table 2 were annealed in pressurized steam [18]. This led to shrinkage (by 24–31%) and, thus, to an increase in denier (by 14–16%). Also, as expected, measurement of the sonic modulus molecular orientation factor showed that the annealed yarns had lower orientation than did the unannealed ones: for cascade stretches of 2X and 7X, the orientation factor was, respectively, 0.59 and 0.65 for the annealed and 0.69 and 0.79 for the other. The results given in Table 5 show that annealing led to an increase in μ. The values of a and n of the annealed yarns, given in Table

TABLE 5 Values of μ for Annealed and Unannealed Acrylic Yarns at Equal Values of Initial Tension (Line Contact; Cascade Stretch 2X)

T_0/r (g_f/mm)	μ	
	Unannealed	Annealed
200	0.16	0.19
300	0.12	0.14

Source: Ref. 13.

TABLE 6 Results for Annealed Acrylic Yarns with Cascade Stretch of 2X

	a $(g_f/mm)^{1-n}$	n
Point contact	0.32	0.86
Line contact	0.45	0.68

Source: Ref. 13.

6, when compared with those of unannealed, given in Table 4, show that the process used in annealing did not cause a change in n but led to an increase (significant in the case of line contact) in a. It is speculated that the increases in μ and a with annealing resulted from the latter causing an increase in diameter which led to an increase in m and/or a decrease in K. The value of n remaining essentially unchanged indicates that the treatment involved did not affect the shape of the pressure–area curve.

5. Effect of the Mode of Contact

As is clear from the results presented in the preceding sections, the line-contact method, as compared to the point contact, invariably led to a higher coefficient of friction. This is an outcome as expected from the fact that the line-contact method involved a larger area of contact (larger m) than did the point-contact method. The data given in Tables 4 and 6 show that generally the value of the constant a was greater and that of the index n smaller for the line than for the point. It is expected that the former result was due to the influence largely of m on a and the latter is due to the difference in the compressional mechanical behaviors obtained in the two modes.

A practical point to be made of the result that line contact led to higher μ than did point contact was that staple length, helical twist, and the orientation of morphological lobes on fibers, or flutes on surfaces of guides, could have significant impacts on the frictional forces generated. The larger the staple length, the smaller the twist in assembly, and/or the larger the orientation of flutes parallel to the direction of sliding, generally the greater was the frictional resistance that could be expected. The accompanying effects of these on the fiber and the machine surfaces would be greater alignment, control, and/or abrasion of fibers in an assembly, and greater wear of the surfaces of contacting parts.

6. Effect of Fiber Types

Two materials on which controlled experiments were conducted and whose frictional properties could be compared under equivalent conditions were polypropylene and acrylic. The results obtained on these materials are given in Table 7.

TABLE 7 Values of Frictional Parameters of Acrylic and
Polypropylene Yarns at Equal Values of T_0/r

	μ ($T_0/r = 200$ g$_f$/mm)	a (g$_f$/mm)$^{1-n}$	n
Polypropylene	0.37	0.59	0.85
Acrylic	0.16	0.32	0.68

Source: Ref. 13.

The values of all three parameters (i.e., μ, a, and n) were higher for polypropylene than acrylic. These differences could arise from the differences that existed in the chemical nature as well as in the physical structures of the two materials. A higher value of n for polypropylene indicates that the behavior in compression was more plastic in nature for this material than it was for the other. Several factors could account for the differences in the values of a and, through a and n, in the values of μ [Eq. (13)]: the strength of the junctions (S), the deformational behavior of the junctions (K, n), and the nature and the size of the contacts during friction tests (C_m, m). With the value of C_m remaining relatively constant and that of n higher, it is obvious that the difference in a arose from polypropylene having a higher S, lower K, and/or higher m than acrylic. Which one or more of these possibilities accounted for the difference could not be ascertained, as independent tests for measuring the values of these microscopic properties did not exist. In an attempt to understand the difference in the frictional behaviors of the two materials, however, several hypothetical combinations of the properties were considered and the pressure–area curves computed. In all cases, it was found that the combinations used led to a greater area of contact and a higher μ in polypropylene than in acrylic [13].

7. Summary

The effects of a variety of factors on the values of μ, a, and n have, thus, been examined. The availability of the structural models for these parameters allowed one to speculate on the underlying causes for the changes in values. For example, one could always construct hypothetical pressure–area curves by using the calculated values of a and n, plugging these along with the assumed values for the constants S, C_m, and m in Eqs. (8) and (9) and calculating the values of K and α. With the values of the two parameters known, Eq. (4) could be plotted. An example is given in Fig. 6, in which the effect of annealing in acrylic yarns was examined. The scenarios selected indicate that for a given set of conditions, the area of contact (i.e., the area corresponding to the point where an isoload curve intersected with the pressure–area curve) was relatively greater for the annealed than the unannealed yarns.

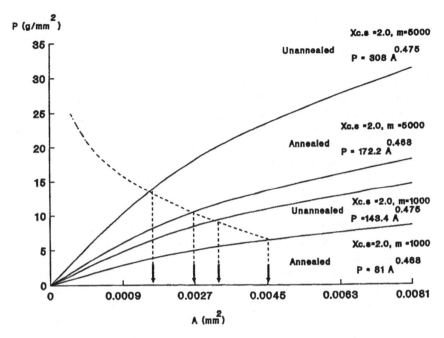

FIG. 6 Hypothetical pressure–area curves showing the effect of annealing in acrylic yarns. ($S = 1$ g$_f$/mm^2, $C_m = 1$). (From Ref. 13.)

IV. FRICTIONAL BEHAVIOR OF HUMAN HAIR

One of the most important physical properties of hair is interfiber friction, which influences several of the hair assembly characteristics, some of these being combability, manageability, hand, and appearance. Most hair aids, in addition to performing the function for which they are designed, affect assembly properties by modifying fiber surface and, through it, friction.

Chemically, hair is a keratin fiber like wool and contains polypeptide chains with about 20 amino acid side groups. The most important component of these is the sulfur-containing cystine group that causes disulfide cross-links to form between chains. These bonds contribute greatly to the physical and mechanical stability of hair. Morphologically, the fiber consists of three components: (1) the cuticle, which is the outermost protective covering, is built of six to eight layers of overlapping scales and contains highly cross-linked amorphous matrix of polypeptide chains; (2) the cortex, which is the bulk of the fiber, contains spindle-shaped crystalline microfibrils and provides much of the mechanical stability and strength; and (3) the medulla, which is the innermost component and is a hollow lumen along the center. Obviously, the component of most interest from the standpoint of hair performance and treatments is the cuticle, which has unique morphology and, therefore, frictional characteristics. The scales which are 0.5 μm thick, 5 μm wide, cover about three-fourths of the circumference, and point from root to tip. This causes the frictional properties to be directional [i.e., friction is higher when the scales engage and oppose each other (against scale friction) and lower when scales slip over each other (with scale friction)]. The cuticle is sulfur rich and contains disulfide cross-linkages; these are, however, susceptible to damage from chemicals used in cosmetic treatments and from sunlight, environmental pollutants and chlorine found in swimming pools.

These voluntary or involuntary exposures affect hair characteristics such as combability, manageability, and so forth by modifying the surface through chemical changes of the outer layers. The studies have generally indicated that the effects are cumulative, especially if the surface is chemically altered by a treatment. Swimming and use of cosmetic treatments being highly popular, the effects of major concern are those due to the exposure of fiber to chlorine, found in pool water, and to chemical compounds, present in bleaching, dyeing, and permanent waving lotions. A study was undertaken which explored the effects of these factors on hair morphology and friction. Two sets of experiments were conducted. In one, the effects of the concentration of chlorine in solution, the number of hours of treatment, and the pH of the solution was explored. In the other, the effects of combining chlorination with cosmetic treatments were examined. Full details of the study can be found in a doctoral thesis [21] and a number of publications [16,22–24]. Brief details of some of the experiments conducted and the results obtained are, however, included here.

A. Experimental Details

1. Materials and Treatments

I.   Dark brown Caucasian hair was Soxhlet extracted using chloroform/methanol mixture. Chlorine solution was prepared by dilution of sodium hypochlorite solution. The pH of the solution was varied and controlled with HCl. The treatments were carried out at 20°C using a 500 mL liquor:1 g hair ratio. In one of the two sets of experiments performed, the pH level was maintained constant at 8 and the tresses were treated with 20, 40, and 60 cycles in solutions containing 0, 10, and 50 parts per million (ppm) concentration of chlorine. In the other set, the concentration was fixed at 50 ppm, but the fibers were subjected to 10 and 20 cycles of chlorination in solutions maintained at pH levels of 8, 4, and 2. Each cycle consisted of soaking a tress for 1 h in treatment bath and drying in an air-circulating oven for 1 h at 40–50°C.

II. Natural blond hair was subjected to chlorination combined with a cosmetic treatment (oxidative dyeing, hydrogen peroxide bleaching, or setting using a commercial permanent waving lotion). These treatments were given either before chlorination (pretreatment) or after chlorination (posttreatment). The concentration of chlorine used was 10 ppm and the number of cycles of treatment administered were 0, 5, 10, 15, and 30.

2. Friction Measurements

The technique employed was the twist method discussed earlier. Measurements were made with two single hair fibers twisted together by two turns of twist, one end of each of which was subjected to a constant tension and the other held in Instron load cell. The fibers were traversed at 0.5 in./min against an initial tension of 3 g_f imposed on each end. From the stick–slip profiles, the values of the static (μ_s) and the kinetic (μ_k) coefficients of friction were determined. Also, the tests were conducted in both the with (μ_w) and the against (μ_a) scale directions. From these, the values of the differential frictional effect (DFE), given by $\mu_a - \mu_w$, was determined using the static values.

3. Morphology

In order to understand the nature of the changes brought about by the treatments in hair friction, changes in morphology were examined under SEM. The portions of fibers examined were those where rubbing had actually taken place during the friction tests.

B. Results

1. Effect of Chlorine in Solution

The results illustrated in Fig. 7 show that the first increment in concentration (0–10 ppm) led to significant increases in μ in both the "with" and the "against" scale values. Further increase in concentration, however, did not produce significant change. On the effect of the cycle also, it is seen that the greatest increase in the value occurred by about 20 cycles; a further increase in chlorination cycles caused little or no change in μ.

A change in hydrogen-ion concentration greatly affected coefficient of friction. Hair chlorinated in acidic region had a substantially higher value of μ than the one treated in neutral region. A greater change was noted in μ with pH at lower cycles (10) than at higher (20) and a greater change occurred in transition from pH 8 to 4 than pH 4 to 2.

The results given in Table 8 indicate that DFE decreased substantially with increase in concentration (ppm). This is due to the destruction of scales, illustrated later. The effect the cycles produced varied with the concentration. At 0 ppm concentration, cycles from 0 (control) to 40 caused an increase in DFE and then

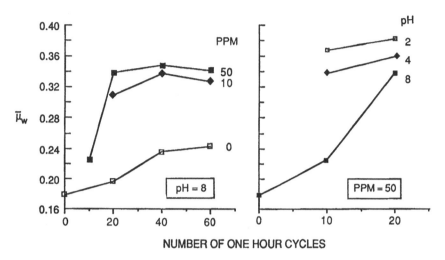

FIG. 7 Effect of chlorination cycles at different concentrations and pH on the average coefficient of friction. (From Ref. 16.)

little further change in it with an increase in cycles to 60. This increase in DFE was expected from the scales becoming more prominent with soaking, swelling, and drying. At 10 ppm, the DFE was already reduced to half or less of the value at 0 ppm, and it further decreased but only by a small amount with increase in cycles. At 50 ppm, the value which was already low at 20 cycles changed inconsistently with increase in cycles.

The values of DFE (static) given in Table 9 show that with an increase in acidity, a marked decrease occurred at 10 cycles. At higher cycles [20], the value which was already low at neutral pH changed little with increase in acidity.

An important property of viscoelastic fibers was that under suitable conditions of testing, they produced stick–slip profile with $\mu_s > \mu_k$. Because a chemical treatment affected not only the morphology but also the mechanical properties, the effect of a treatment could be expected to appear more sensitively on the nature of a stick–slip profile than on an average value of μ.

Typical examples are shown in Fig. 8, which illustrate the effects of cycles and pH. The difference in stick and slip values generally increased with an increase in the cycles, the concentration of chlorine, and the acidity of the solution. Because all friction tests were performed with a constant speed of traverse, constant initial tension, and a constant number of turns of twist, the force profiles reflect a change in the state of the surface (i.e., from a hard and elastic of control and mildly chlorinated specimens to a softened and plastic surface of severely chlorinated materials. Similar results have been reported on the chlorination of wool [25,26].

TABLE 8 Effect of Cycles of Treatment and Concentration of Chlorine on DFE ($\times 10^2$) at pH 8

	Cycles									
	0	20			40			60		
ppm	0	0	10	50	0	10	50	0	10	50
DFE	6.5	8.8	4.3	1.6	9.4	3.9	3.0	9.5	3.2	1.9

TABLE 9 Effect of Cycles and pH on DFE ($\times 10^2$) at 50 ppm

			Cycles				
	0		10		20		
pH	8	8	4	2	8	4	2
DFE	6.5	5.7	3.9	2.7	1.6	3.0	1.0

Disulfide bond scission and peptide link cleavage [27] are expected to occur in hair and affect the surface just as reported to occur in wool.

The changes in the surface morphology and frictional properties alluded to earlier could be verified by examination of fiber surfaces where rubbing had actually occurred during friction tests. Selected examples of the effect of treatment under neutral conditions are given in Fig. 9. The 60 cycles control shows definitive scale structure and damage due to rubbing minimal (Fig. 9a). Treatment at 10 ppm and 60 cycles led to a significant loss of scale definition (Fig. 9b). The increase in concentration to 50 ppm resulted in a greatly softened surface with bulk deformation due to rubbing (Fig. 9c). Increasing the acidity of the solution hightened the damage; under highly acidic conditions, the dissolution combined with frictional damage was enough in some cases to reveal the underlying cortex [16].

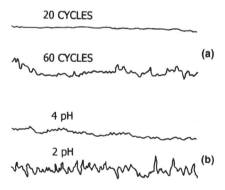

FIG. 8 Effect of chlorination on force profiles. (a) Effect of cycles at 10 ppm and pH 8; (b) effect of pH at 20 cycles and 50 ppm. (From Ref. 21.)

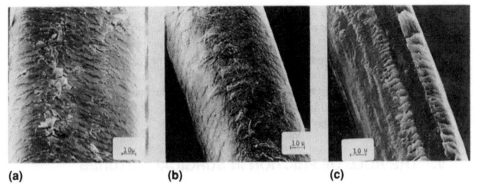

(a)   **(b)**   **(c)**

FIG. 9 Scanning electron micrographs of chlorinated and friction tested dark brown hair fibers: (a) control, 60 cycles at 0 ppm and pH 8, "against" scale; (b) 60 cycles at 10 ppm and pH 8, "with" scale; (c) 60 cycles at 50 ppm and pH 8, "against" scale. (From Ref. 21.)

2. Effect of Combining Chlorination with Cosmetic Treatment

The effect of cosmetic treatment sequence and number of 1-h cycles of chlorination of $\bar{\mu}$ are given in Table 10. Treatment and cycle effects are evident. Cosmetic treatments alone significantly increased the coefficient of friction beyond that of the untreated control (0 cycles). When chlorination was added to cosmetically

TABLE 10 Effect of Cosmetic Treatment Sequence and Cycles of Chlorination on Average μ in the with-Scale Rubbing

		Cycles				
		0	5	10	15	30
Pre-treatment						
	Control	0.13	0.18	0.22	0.22	0.30
	Bleach	0.19	0.27	0.27	0.31	0.31
	Dye	0.18	0.22	0.28	0.25	0.36
	Perm	0.20	0.19	0.21	0.25	0.31
Post-treatment						
	Control	0.13	0.18	0.22	0.22	0.30
	Bleach	0.19	0.25	0.26	0.29	0.25
	Dye	0.18	0.18	0.22	0.21	0.21
	Perm	0.20	0.25	0.26	0.31	0.27

Source: Ref. 24.

treated specimens, either after or before, in all cases except one, the coefficients of friction increased. In the posttreated specimens, the value of μ tended to decrease somewhat after 15 cycles. The exception noted was the postcolored specimens in which case the treatment did not cause additional changes in μ. Examination of surfaces under SEM showed that in the posttreated materials with cycles above 15, the surface tended to be smooth and hard. This indicated that the layers softened by prior chlorination were sloughed off by cosmetic treatments exposing underlying unaffected layers with lower friction.

V. THE ROLE OF FRICTION IN SURGICAL SUTURES

Following surgery, a wound is usually closed with a suture thread, which is looped around the cut or repaired vessels and held in place with a square knot. The latter consists of the mechanical interlacing of the ends and resists untying or slippage when subjected to tension because of the frictional restraints generated within the structure. The frictional force between the elements of the knot depends on the coefficient of friction, the number of crossing points, the angle of contact between the threads at each crossing point, and the normal force pressing the threads against each other. The normal force is initially governed by the tension used in tying a knot and later by the compressed tissues in an effort to expand and regain their original sizes. Clearly, if the total frictional force is high, the knot may hold firmly; if not, the components of it may slide and result in failure. Thus, an understanding of the effects the material and the structure of the suture and the factors used in constructing knots had on suture friction was important. Studies were undertaken to develop such an understanding.

A. Methods

A knot consists of a number of throws pressed against each other and held by frictional contacts. The method that closely simulates the construction of a surgical knot is the twist method. Using a simple analogy, each turn of twist in the method could be taken to correspond to one throw (Fig. 10). T_0 could be considered as the tension at the entrance to or the exit from the knot where the ends were gripped by compressive forces, and T as the tension in the suture loop, exerted by the compressed tissues, necessary for the knot to fail by slippage. One could consider the knot to be secure if the tension actually existing in the loop (T_a) was less than T [28].

Because there was a wide range of tensions the surgeons used in tying knots, the measurement of μ was carried out at different values of T_0. The number of throws used in one detailed study [17] was three which simulated a three-throw knot. The suture materials used were the following: Dexon®, a polyglycolic acid

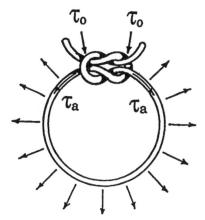

FIG. 10 Model of a surgical knot. (From Ref. 28.)

braided multifilament yarn; Mersilene®, a polyester braided multifilament yarn; Tevdek II®, a polyester braided and Teflon-impregnated multifilament yarn; Ticron®, a polyester braided and silicon-treated multifilament yarn; silk, a braided yarn; Ethilon®, a nylon monofilament yarn; Surgilon®, a nylon braided and silicon-treated yarn; and Prolene®, a polypropylene monofilament yarn.

B. Results

Differences among sutures were obvious in terms of both the average values of μ (Fig. 11) as well as the nature of the stick–slip profiles (Fig. 12). All materials were approximately of the same diameter and, therefore, had the same T_0/r value at any given value of T_0 [Eq. (13)]. The value of μ decreased, as expected, with an increase in the applied tension. The two monofilament materials (polypropylene and nylon) had the highest value of μ at the lowest tension (0.125 lbs) and ended up having the lowest value of μ at the highest tension. The effect of geometric configuration could be seen by comparing results for Surgilon and Ethilon, one braided and the other monofilament. Although the former was silicon treated, its μ was higher than that of the latter. The effect of surface modification could be examined by comparing the results of Mersilene, Ticron, and Tevdek, all polyester materials and braided but differing in terms of the treatment given to the surface. The ones with special finishes gave consistently lower values. The knot-holding force, the force in tensile tests at which a knot failed by slippage, if it was not secure, or by rupture, if it was secure, measured on the three materials gave the results as shown in Fig. 13 [29]. Clearly, special coatings given to Ticron

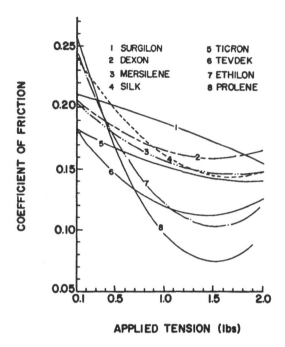

FIG. 11   Frictional behavior of surgical sutures. (From Ref. 17.)

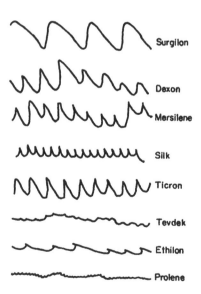

FIG. 12   Stick–slip profiles of sutures; size 2/0, $n = 3$, $T_0 = 454$ g_f. (From Ref. 17.)

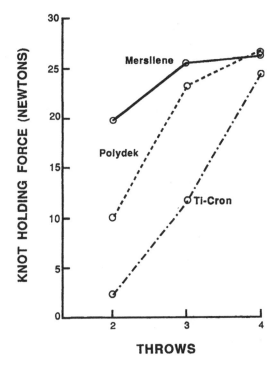

FIG. 13 Effect of throws on knot-holding force in dry tests, size 2/0. (From Ref. 29.)

and Polydek led to much lower values of knot-holding force of the materials at two throws, and in the case of Ticron also at three throws, than found on uncoated Mersilene. The three sutures having the same chemical constitution and physical size, and constructions gave, as expected, approximately the same knot-holding force values when the knot was made fully secure by using four throws.

As pointed out earlier, viscoelastic materials, of which sutures were made, produced friction profiles which were stick–slip in nature. Results in Fig. 12 show that different sutures gave different profiles which differed widely in terms of the amplitude and the period of vibration. Surgilon gave the highest amplitude and Tevdek, Ethilon, and Prolene the lowest amplitudes. Although most suture materials produced a regular pattern, Prolene gave a composite structure with secondary fluctuations superimposed over primary. A quantitative measure of stick–slip was the difference $\mu_s - \mu_k$. In clinical work, this parameter could be important, as it characterized the potential a knot will hold again if it slipped at some point. It was expected that in instances in which $\mu_s - \mu_k$ was small, a suture once beginning to slip will most likely continue to slip with less tendency to

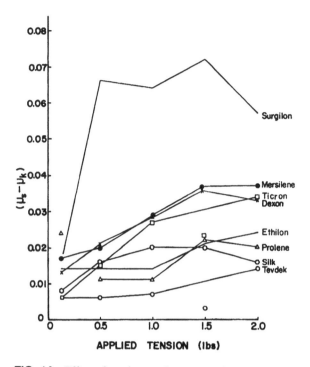

FIG. 14   Effect of tension on the $\mu_s-\mu_k$ values. (From Ref. 17.)

rehold. The value of this parameter is plotted against T_0 in Fig. 14 which shows that the value increased with T_0.

The results in this section, thus, show that frictional properties were important in determining the success of sutures in surgery, as these properties affected not only the handling and suturability characteristics of the threads but also the security of the knots tied in them.

VI.   FRICTIONAL BEHAVIOR OF FABRICS

Frictional properties of textile fabrics affect aesthetics and comfort in wearing. Surface forces associated with fabrics also play critical roles in influencing handling, transport, and other processes involved in converting a fabric into final products. Increased emphasis on speed of processing and automation in apparel manufacturing in recent years has made it necessary that a fabric-to-fabric and fabric-to-other surfaces frictional responses be understood.

Briefly described in this section are the results from an investigation in which the frictional properties of woven fabrics were assessed [30,31]. Two sets of fabrics were considered. In one, commercial fabrics ranging broadly in terms of fiber type, yarn structure, fabric weight, and weave construction were selected. In the other, a set of model fabrics varying systematically in terms of yarn twist and weave tightness were produced and tested. The main operational variable used in both cases was the contact pressure.

A. Characterization of Behavior

Because fabrics were two-dimensional structures, contact usually involves large but measurable geometric areas. Accordingly, Eq. (3) was modified to characterize interfabric friction by taking ratios of the frictional force, F, and the normal force, N, to the geometric area of contact, A' [32]:

$$\frac{F}{A'} = a'\left(\frac{N}{A'}\right)^{n'}$$

$$\ln\left(\frac{F}{A'}\right) = \ln(a') + n'\ln\left(\frac{N}{A'}\right)$$

In this, A' is the geometric area of contact, and a' and n' are the constants for a fabric which correspond respectively to the constants a and n, used earlier for one-dimensional structures.

B. Experimental

In the first phase, 26 fabrics used commercially were selected. These included 10 cotton, 3 rayon, 5 polyester, 2 linen, 1 wool, 1 silk, and 4 cotton/polyester blended materials. Three of the materials were woven from continuous filament yarns (two rayon and one polyester), and some of the fabrics were in both the griege and the finished (durable press, mercerized, heat set) states.

For the study in the second phase, a series of plain woven fabrics varying systematically in yarn denier, twist constant, and weave tightness was prepared using polyester yarns on a narrow loom.

The device used for conducting tests is shown in Fig. 15 [31]. It consists of a crosshead, a horizontal metal plate, a top square block, and a digital force gauge. A piece of fabric (25.4 cm × 15.2 cm) is placed flat and clamped securely on the bottom plate. The top block, with a contact area of 6.45 cm², is also covered with fabric and connected to the force gauge. When the crosshead is run and the plate is traversed, the top block resists the motion; the force so generated is re-

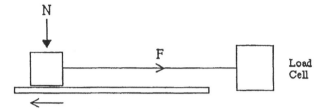

FIG. 15 Schematic of the device used for conducting tests on fabrics. (From Ref. 31.)

corded by the gauge. Tests were conducted at a given speed but with several different normal pressures.

C. Results

By properly aligning the specimens, tests were conducted with the filling threads sliding parallel to the filling (filling–filling). The values of $\mu(N')$ varied greatly among the fabrics and covered a broad range (0.1–0.9). The upper and lower bounds of this range are shown in Fig. 16 [30]. It is noted that the spread in values varied with the normal pressure; the spread, which is wide at low pressure, decreased as pressure increased. This was an interesting result and indicated that

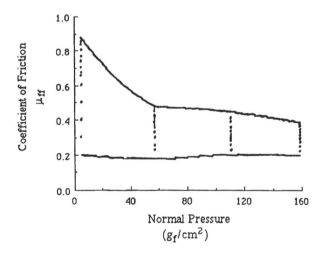

FIG. 16 Range of the coefficient of friction values found on commercial fabrics. (From Ref. 30.)

at low pressures, such as encountered in ply separation and fabric transport, different fabrics in terms of their frictional behavior different greatly from each other, whereas at high pressure, such as encountered in sewing, the differences were not perceived to be as great.

Among other results, it was noted that fabrics given any type of a finishing treatment usually had a higher coefficient of friction than unfinished griege material at all levels of pressure. Whether this was due to the tacky nature of adhesion characteristic of a topically applied and cured finish or to an increase in the area of contact arising from the smoothening of the surface was not fully understood. Another interesting result noted was that fabrics made from continuous filament yarns showed distinctly different frictional properties than did the fabrics made from spun yarns. The former gave lower coefficients of friction (often as low as 0.1), higher n' values (sometimes nearly 1) and lower a' values than did the latter. A plot of a' against n' given in Fig. 17 shows that the two fabric groups occupied distinctly different positions. These results also show that the two friction indices, although estimated independently, were intimately related. Such relationships could be generally as expected if the model developed on fibers [Eq. (9)] also applied to fabrics and the values of the parameters S, C_m, K, and m either did not vary much from one fabric to another or they changed in a manner that their combined effects did not alter the primary (inverse) relationship.

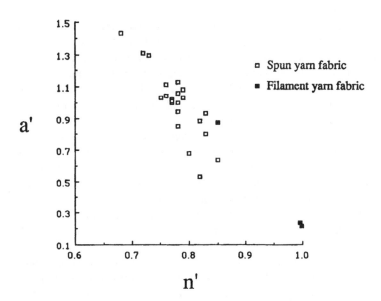

FIG. 17 Plot of a' against n'. (From Ref. 31.)

$$a = SC_m K^{-n} m^{1-n} \qquad\qquad (9)$$

Because the fabrics studied here are commercial, a systematic variation did not exist in their structures. Therefore, the effects of the material and the construction factors on the values of the indices a' and n' could not be determined. Accordingly, in the second phase of the investigation, model fabrics were prepared which allowed an examination of some of these effects.

Full details of the study on model fabrics can be found in Ref. 31. Briefly, the results obtained showed that yarn twist constant and weave tightness affected fabric frictional properties significantly. The value of the friction parameter a' decreased and that of n' increased with an increase in the twist constant and in the weave tightness. It was shown that the changes in the values of these parameters with the changes in the structure could be rationalized by considering the effect the latter produced on fabric compressibility and relative area of contact.

VII. CONCLUDING REMARKS

Although perhaps several mechanisms govern the frictional phenomenon in fibrous materials, the one that has served well in explaining the results in general is the adhesion-shearing of Bowden and Tabor. The frictional force is given by the product of the specific shear strength of the junctions and the true area of contact. The latter is determined by the pressure imposed and the mechanical properties of the contacting asperities in compression. Using a generalized model for the pressure–area relation, detailed structures are given to the two friction indices, a and n, and the coefficient of friction, μ. Effects of the chemical nature, the physical structure, and the cross-sectional size and shape of a fiber, the environmental conditions and the treatments to which a material is subjected, and the manner in which the tests are conducted can be rationalized in view of the model. In limited tests conducted, the usefulness of the model in explaining the results on fabrics has also been illustrated.

Clearly, friction is an important physical property of fibrous materials, the understanding of which is essential for both developing optimum processing conditions for manufacturing a product as well as for predicting and controlling the performance of the latter. Although a great deal has been learned about the nature of friction and its influence on the behavior of the processes and products, further work is needed to close gaps in and enhance our understanding of the phenomena. For example, work is needed to determine the relation friction may have with other surface properties, in particular, surface energy. It is possible that at some pressure (e.g. low), the magnitude of the surface energy of the contacting bodies and the level of compatibility between the two in terms of the polar and nonpolar values determines the adhesion and, therefore, the frictional resistance. One of the parameters of the structural model for which there is no direct measure is the specific shear strength of the junctions. If the connection exists, it is feasible that a value for this can be derived from the considerations of the surface energies of the materials.

Second, it has been observed that often the effect of a treatment is more prominent on the nature of the stick–slip profile than on the average value of the coefficient of friction. Understanding of the exact mechanism that gives rise to the stick–slip curve is lacking but is needed in order to fully appreciate the concept and advance our knowledge of the field.

Third, much of the theoretical foundation centers around the nature of the pressure–area relationship that exists between two contacting bodies. A direct measure for this is not available but is needed in order to confirm or modify the prevailing view. If the results show that the current concept is sound, then it should be possible to assess directly the values of the parameters K and n of the model and determine how these vary with a material's chemical and physical structure. If, on the other hand, it is found that the present view was not exactly valid but needed to be modified, a general direction for this could become available from the results.

Finally, because textile products are built of fibers and yarns, it is proposed that structural models correlating frictional properties of an assembly with those of the constituent elements be developed which will provide scientific tools for engineering products with desired aesthetic and functional properties.

ACKNOWLEDGMENTS

This chapter is based on the works of several of the writer's colleagues and graduate students. The writer thanks the following individuals: Dr. Timothy G. Clapp, Dr. Nancy Fair, Dr. Yehia El-Mogahzy, Dr. Ramond W. Postlethwait, Dr. Neelesh B. Timble, and Dr. Kay W. Wolf.

REFERENCES

1. F. P. Bowden, and D. Tabor, *The Friction and Lubrication of Solids*, Oxford University Press, London, 1950.
2. F. P. Bowden and D. Tabor, *The Friction and Lubrication of Solids, Part II*, Oxford, University Press, London, 1964.
3. J. Lindberg and N. Gralen. Text. Res. J. *18*:287 (1948).
4. F. P. Bowden and J. E. Young. Proc. Roy. Soc. *A208*:444 1951.
5. N. Gralen, B. Olofsson, and J. Lindberg. Text. Res. J. *23*:623 (1953).
6. H. G. Howell. Text. Res. J. *23*(8):589 (1953).
7. H. G. Howell and J. Mazur. J. Text. Inst. *44*:T59 (1953).
8. B. Lincoln. Br. J. Appl. Phys. *3*:260 (1952).
9. A. S. Lodge and H. G. Howell. Proc. Phys. Soc. London *B67*:89 (1954).

10. J. Mazur. J. Text. Inst. *46*:T712 (1955).
11. A. Viswanathan. J. Text. Inst. *57*:T30 (1966).
12. B. S. Gupta and Y. E. El-Mogahzy. Text. Res. J. *61*(9):547 (1991).
13. Y. E. El-Mogahzy and B. S. Gupta. Text. Res. J. *63*:219 (1993).
14. T. Nogai, Y. Norumi, and M. Ihara. J. Text. Machinery Soc. Japan *21*(2):40 (1975).
15. M. J. Schick. Text. Res. J. *50*:675 (1980).
16. N. Fair and B. S. Gupta. J. Soc. Cosmet. Chem. *33*:229 (1982).
17. B. S. Gupta, K. W. Wolf, and R. W. Postlethwait. Surg. Gynecol. Obstet. *161*:12 (1985).
18. Y. El-Mogahzy. Ph.D. dissertation, North Carolina State University, Raleigh, (1987).
19. H. G. Howell. J. Text. Inst. *45*:T575 (1954).
20. B. S. Gupta, Y. E. El-Mogahzy, and D. Selivansky. J. Appl. Polym. Sci. *38*:899 (1989).
21. N. B. Fair. Ph.D. dissertation, North Carolina State University, Raleigh (1984).
22. N. B. Fair and B. S. Gupta. J. Soc. Cosmet. Chem. *38*:359 (1987).
23. N. B. Fair and B. S. Gupta. J. Soc. Cosmet. Chem. *38*:371 (1987).
24. N. B. Fair and B. S. Gupta. J. Soc. Cosmet. Chem. *39*:93 (1988).
25. J. H. Bradbury. Text. Res. J. *31*:735 1961.
26. K. R. Makinson and I. C. Watt. Text. Res. J. *42*:698 (1972).
27. K. R. Makinson. Text. Res. J. *44*:856 (1974).
28. B. S. Gupta and R. W. Postlethwait. in *Biomaterials 1980* (G. Winter, D. Gibbons, and H. Plenk, Jr., eds.), Wiley–Interscience, New York, 1982, pp. 661–668.
29. B. S. Gupta. *On the Performance of Surgical Knots in Sutures*, ASME, New York, 1989.
30. T. G. Clapp, N. B. Timble, and B. S. Gupta. J. Appl. Poly. Sci. *47*:373 (1991).
31. N. B. Timble. Ph.D. dissertation, North Carolina State University, Raleigh (1993).
32. D. Wilson. J. Text. Inst. *54*:T143 (1963).
33. W. W. Carr, J. E. Posey, and W. C. Tincher. Text. Res. J. *58*:129 (1988).
34. H. G. Howell. J. Text. Inst. *44*:T359 (1953).
35. M. Ohsawa and S. Namiki. J. Text. Machinery Soc. Japan *12*(5):197 (1966).
36. K. W. Wolf. Master of Science thesis, North Carolina State University, Raleigh (1979).
37. W. Zurek, D. Jankowiak and I. Frydrych. Text. Res. J. *55*:113 (1985).

4
Infrared Absorption Characteristics of Fabrics

WALLACE W. CARR and ELIZABETH G. McFARLAND School of Textile and Fiber Engineering, Georgia Institute of Technology, Atlanta, Georgia

D. S. SARMA Trident, Inc., Brookfield, Connecticut

I. INTRODUCTION

For many years, the textile industry has used infrared (IR) radiation for predrying, drying, and curing of various textile products. Until recently, most of the heating applications have used IR sources with emissions in the medium to long infrared regions of the spectrum. Improvements in electric-based IR equipment, particularly in emitter materials and controls, have allowed the adjustment of the region (short, medium, or long wavelength) in which peak emission occurs. Because there are several types of IR emitters available for heating textiles, the type of emitter that should be used for a given application is often an issue. The interaction of IR energy with fabrics and the effect of emitter type on the efficiency of heating fabrics is discussed in this chapter. The material presented here is based on research conducted at Georgia Tech [1,2]. Although the efficiency of IR heating may depend on factors such as oven design and the interaction of convective and conductive modes of heat transfer with infrared heating, only the interaction of IR energy with fabrics is considered in this chapter.

Several experimental studies on the interaction of IR emitters with fabrics have been reported [3–9]. Average absorptivities of fabrics heated with a tungsten source at 3160 K were measured in a textile flammability study [3,4,9]. Most of the data were for undyed woven and knit fabrics, and for cotton, polyester, polyester/cotton, nylon, acetate, and nylon/acetate fibers. The average absorptivities of woven and knit structures were similar; however, the average absorptivity for a cotton pile fabric was appreciably lower than for the woven and knit fabrics. The effect of dyeing on average absorptivities was larger than those reported in Ref. 1, probably due to the differences in sources for the two studies.

Broadbent et al. [5] compared the energetic efficiencies of two types of IR sources used to predry textile fabrics. Energetic efficiency (EFF) was defined as the ratio of power converted into evaporated water to electric power input. The performances of a medium-wavelength quartz tube source and a short-wavelength T-3 emitter were compared. In one set of studies, cotton, polyester, and polyester/cotton fabrics were predried from a moisture regain of approximately 85 to 20%. The EFF of the medium wavelength quartz tubes was approximately 73%, whereas it was about 56% for the T-3 tubes. A lower EFF (48.6%) was obtained when the medium-wavelength quartz tubes were used to predry a plain woven nylon fabric of similar weight to one of the cotton fabrics. A second set of predrying tests were conducted where the fabrics were vacuumed to reduce the moisture regain to 50–55% for cotton and 40–45% for polyester. The final moisture regain to which the fabrics were dried was lower (as low as 10%). For these tests, EFFs for the quartz and the T-3 tubes were both about 50%.

In a study by Campagna et al. [6], color had little effect on the heating and drying of fabrics for medium-wavelength IR sources. The influence of short-wavelength sources on drying and heating varied with fabric type. Although black-and-white 100% polyester fabrics had identical heating and drying rates, a black 50/50 cotton/polyester fabric heated and dried with greater efficiency than a corresponding white fabric.

II. BACKGROUND

A. Thermal Radiation

Thermal radiation is a mode of heat transfer where energy is transported in the form of electromagnetic waves. Thermal radiation is emitted solely by virtue of a body's temperature. Although extremely high-temperature sources emit thermal radiation in the ultraviolet and visible regions of the spectrum, thermal radiators that are used for industrial heating applications (below 2500 K) emit primarily in the infrared region of the spectrum. Thus, they are called infrared heaters.

The quantity of energy emitted from a surface as radiant energy depends on absolute temperature and the nature of the surface. A blackbody is a standard with which all thermal radiators and all absorbing surfaces can be compared. It emits the maximum amount of energy that is thermodynamically possible at a given temperature and absorbs all incident radiation. Thermal energy emitted from the surface of a body is distributed over the electromagnetic spectrum. Planck's distribution [10] gives the spectral variation of thermal emission of a blackbody. Emissive power, $E_{b\lambda}(T)$, is the power emitted per unit surface area at a given wavelength λ. The total amount of radiant energy per unit area can be found by integrating $E_{b\lambda}(T)$ over all wavelengths. For a blackbody, this is given by the Stefan–Boltzman law as follows:

$$E_b(T) = \sigma T^4 \tag{1}$$

where $E_b(T)$ is the total emissive power or total radiant power per unit area, σ is the Stefan–Boltzman constant, and T is the absolute surface temperature.

The power emitted per unit surface area at a given wavelength λ can be normalized by dividing the spectral emissive power at each wavelength by the total emissive power. The effect of emitter temperature on the spectral distribution of blackbody emissions is shown in Fig. 1. The area under each of the normalized curves is equal to 1. At a high emitter temperature, much of the radiation is concentrated in the short-wavelength region of the spectrum. At a low emitter temperature, the radiation spreads throughout the spectrum with little close to the visible region.

Radiation incident on a material can be absorbed, reflected, and/or transmitted. The following equation gives the relationship between the absorbed, reflected, and transmitted energy at each wavelength (λ):

$$\alpha_\lambda + \rho_\lambda + \tau_\lambda = 1 \tag{2}$$

where α_λ is the spectral absorptivity, the fraction of incident energy absorbed at a given wavelength, ρ_λ is the spectral reflectivity, the fraction of incident energy reflected at a given wavelength, and τ_λ is the spectral transmissivity, the fraction

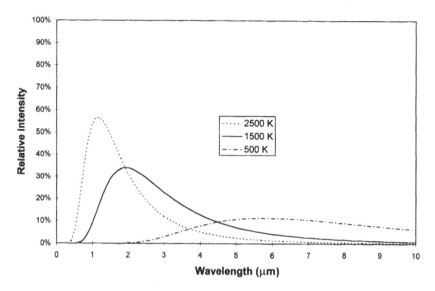

FIG. 1 Normalized blackbody emissions.

of incident energy transmitted at a given wavelength. Normally, ρ_λ and τ_λ are measured, and α_λ is calculated using Eq. (2).

If blackbody emission is incident on a body having a spectral absorptivity of α_λ, the fraction at each wavelength absorbed is the product of α_λ and $E_{b\lambda}(T)/E_b(T)$. By integrating $\alpha_\lambda E_{b\lambda}(T)/E_b(T)$ over all wavelengths, the integrated average absorptivity $(\overline{\alpha})$ is obtained. The integrated average absorptivity is the fraction of the total incident radiation absorbed by the fabric:

$$\overline{\alpha} = \int_0^\infty \alpha_\lambda E_{b\lambda} (T)/E_b (T)d\lambda \tag{3}$$

B. Infrared Emitters

Infrared sources are often classified according to the wavelength at which their peak emission occurs. The peak emissions of short-, medium-, and long-wavelength emitters are less than 1.6 μm, between 1.8 and 3 μm, and above 3 μm, respectively. Some of the characteristics of the types of emitters are quite different [1,11–14]. The short-wavelength emitters have high operating temperatures, typically between 1600°C and 2200°C. Medium-wavelength emitters and long-wavelength emitters operate over temperature ranges of 700–1300°C and 300–700°C, respectively. There are considerable differences in response times and heat fluxes of short- and long-wavelength emitters. Heat fluxes are as high as 150 kW/m^2 for short-wavelength emitters, but are limited to approximately 50 and 20 kW/m^2 for medium- and long-wavelength emitters, respectively. Response times for short-, medium-, and long-wavelength emitters are typically 1 s, 1 min, and several minutes, respectively. Thus, short-wavelength emitters have an advantage in applications where either high heat fluxes or fast response times are needed.

When evaluating the efficiency of IR emitters in heating fabrics, the radiant efficiency (η) of the emitter must be considered as well as integrated average absorptivity. The radiant efficiency (η) of the emitter, a measure of the emitters ability to convert input power (P_{input}) to IR radiant power (P_{IR}), is defined as

$$\eta = \frac{P_{IR}}{P_{input}} \tag{4}$$

The input power to the emitter is lost from the emitter through thermal radiation (P_{IR}), convection ($P_{convection}$), and other losses (P_{losses}):

$$P_{input} = P_{IR} + P_{convection} + P_{losses} \tag{5}$$

As temperature is increased, the power emitted as infrared radiation increases much faster than the power lost through convection. This occurs because thermal radiation (P_{IR}) is proportional to T^4, whereas convection is proportional to T. As a result, short-wavelength (high-temperature) emitters have higher radiant efficiencies. The variation of radiant efficiency of electrical emitters with emitter temperature is shown in Table 1. Although the radiant efficiency (η) varies with emitter design, the values can be considered typical for electrical emitters [13,15]. For example, whereas η for a tungsten lamp is 38% at 500 K, it is as high as 86% at 2500 K. The radiant efficiency is much higher for electrical emitters than gas emitters, because P_{losses} is small for electrical emitters, but is very high for gas emitters due to very hot flue gases leaving the emitter.

III. METHODOLOGY

A. Measurements of Spectral Absorptivities

An integrating sphere coupled with a Fourier transform infrared (FTIR) spectrometer were used to measure the fabric spectral absorptivities reported here. Other investigators have used this technique to measure spectral absorptivities of various materials such as paper, roof tiles, and skin [16–18]. The technique will be briefly discussed, and further details can be found in Refs. 1, 2, 16–18.

TABLE 1 Typical Radiant
Efficiencies of Electrical
Emitters

Emitter temperature	Radiant efficiency
922	43%
1033	46%
1144	49%
1256	52%
1367	57%
1478	62%
1589	66%
1700	70%
1811	73%
1922	76%
2033	78%
2144	80%
2256	82%
2367	84%
2478	85%
2589	87%

Source: Refs. 13 and 15.

As a consequence of geometric and reflecting characteristics of an ideal integrating sphere, radiation incident on the wall of the sphere is reflected uniformly in all directions. As a result, the radiant flux at the wall is uniform and proportional to the total amount of light entering the sphere. A photodetector mounted at a port in the sphere wall can be used to measure the radiant flux. The ratio of the fluxes with and without the sample in the entrance port is a measure of transmissivity. Similarly, the ratio of radiant fluxes with and without the sample in the sample port is a measure of the reflectivity of the sample. Although monochromatic radiation can be used to obtain spectral data, a much faster approach is to use an FTIR spectrometer, which resolves the radiant flux into its spectral components.

Because integrating spheres are not ideal, corrections can be made to improve the accuracy of the measured values of τ_λ and ρ_λ. Ojala et al.'s model [17] provides equations to correct for overfill and substitution as well as the variation in spectral reflectivity of the integrating sphere wall. These model equations were used with data from FTIR interferograms to obtain τ_λ and ρ_λ of the fabrics, and spectral absorptivities were then calculated using Eq. (2)

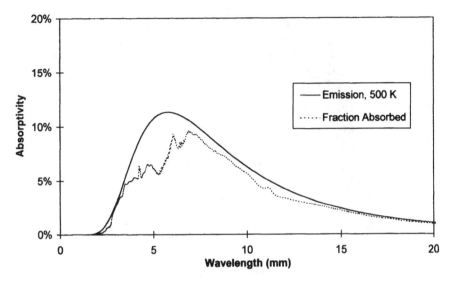

FIG. 2   Average absorptivity for blackbody emission.

B.   Calculation of Integrated Average Absorptivities

Integrated average absorptivities ($\overline{\alpha}$) depend not only on the spectral absorptivities of the fabric but also on the spectral emission of the IR source. Using the spectral absorptivities of the fabrics and blackbody emissions, integrated average absorptivities were calculated for blackbody temperatures ranging from 500 to 3000 K. These calculation will be explained using Figs. 2 and 3. The solid curve represents the normalized spectral emission of a blackbody, and the area under the curve is equal to 1. When the blackbody emission at each wavelength is multiplied by the corresponding fabric spectral absorptivity (illustrated in Fig. 3), the fraction of the incident energy at each wavelength that is absorbed by the fabric is obtained. This fraction is shown as the dashed line in Fig. 2, and the area under this curve is the integrated average absorptivity, the fraction of the total incident radiation absorbed by the fabric. The normalized emission of a blackbody depends on emitter temperature as illustrated in Fig. 1. Thus, integrated average absorptivity also varies with emitter temperature.

C.   Calculation of Overall Radiant Efficiencies

Evaluation of the overall radiant efficiency of IR emitters in heating fabrics must consider emitter radiant efficiency (η) as well as integrated average absorptivity. Overall radiant efficiency (ξ) was calculated by multiplying η (see Table 1) at

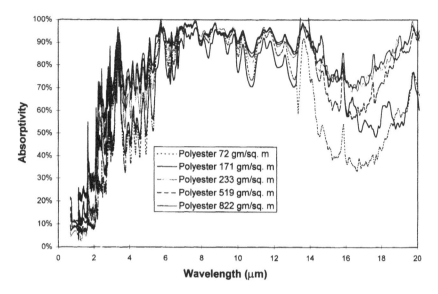

FIG. 3 Spectral absorptivity versus wavelength for different weight polyesters at standard conditions.

a given temperature by the corresponding value of $\overline{\alpha}$. This calculation considers only incident IR energy directly from the emitter and neglects IR energy that is reflected and transmitted by the fabric that may reach the fabric on subsequent reflections. Thus, the heating efficiency of a well-designed IR oven should be higher than ξ if the oven has reflectors and other modes of heat transfer are present [5,16].

IV. FABRIC IR ABSORPTION

A. Parameters Influencing IR Absorption

Several fabric parameters influence spectral absorptivities, average absorptivities, and overall radiant efficiencies. The parameters considered in this chapter include fabric weight (areal density), moisture regain, fiber type, dye, and fabric construction. Because the spectral emissions of the IR emitter are used to calculate integrated average absorptivity and overall radiant efficiency, blackbody emitter temperature is also an important parameter to be discussed. The information presented in this section is based on research conducted at Georgia Tech [1,2].

B. Spectral Characteristics

1. Fabric Weight (Areal Density)

The effect of fabric weight (areal density) on spectral absorptivity of fabrics is illustrated in Fig. 3. The lower-weight fabrics have significantly lower infrared absorption over much of the spectrum. As fabric weight is increased, spectral absorptivity increases; however, a threshold is reached where further increase in fabric weight has little effect. Spectral absorptivity in the near-infrared region (NIR) is low for all fabric weights.

Fabric weight affects both spectral transmissivity and reflectivity (see Figs. 4 and 5), but the changes in spectral transmissivity are much larger. As the fabric weight is increased, the spectral transmissivity decreases rapidly and approaches zero for the heavier fabrics through much of the IR spectrum. However, spectral transmissivity is nonzero in the NIR even for the heavy-weight fabrics (20% for polyester fabric weighing 882 g/m^2). As the fabric weight is increased, the spectral reflectivity increases, but does not increase as rapidly as the spectral transmissivity decreases. Even for heavy-weight fabrics, spectral reflectivity is less than 30–40% throughout the spectrum, except in the NIR. For wavelengths below 2 µm, spectral reflectivity increases significantly with increasing fabric weight and

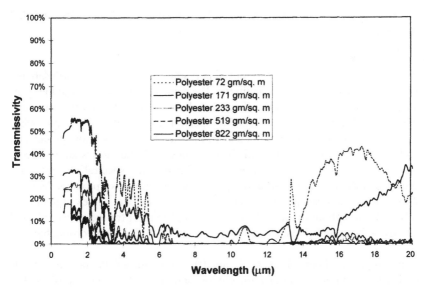

FIG. 4 Spectral transmissivity versus wavelength for different weight polyesters at standard conditions.

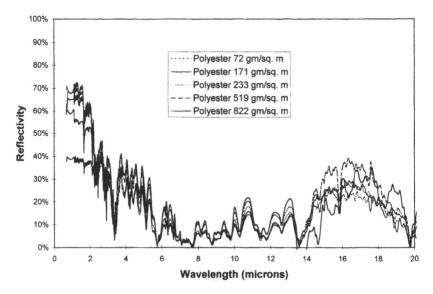

FIG. 5 Spectral reflectivity versus wavelength for different weight polyesters at standard conditions.

is approximately 75% for cotton and polyester fabrics weighing 254 and 822 g/m², respectively.

2. Fiber Type

In Figs. 6–8, the spectral absorptivities of polyamide, cellulosic, and hydrophobic fabrics are shown for standard conditions. The general trends throughout the spectrum are similar for all the fabrics; however, there are some local differences due to characteristic features of the fibers. The spectral absorptivities of the hydrophobic fibers tend to be lower and fluctuate more than those of the polyamides and cellulosics. In the NIR, the spectral absorptivities of all the fabrics are relatively low. As the wavelength is increased above 2.5 μm, the spectral absorptivities increase rapidly, peaking at around 3 μm. Then, spectral absorptivities decrease before reaching a local minimum between 4 and 6 μm. The values remain relatively high throughout the rest of the spectrum of interest for the polyamide and cellulosic fabrics, but fluctuate between 50% and 90% for the hydrophobic fabrics.

The polyamide fibers (Figs. 6, 9, and 10) and cellulosic fibers (Figs. 7, 11, and 12) have similar spectral absorptivity, and spectral reflectivity plots. For wavelengths greater than approximately 2.5 μm, wool has higher spectral absorptivities than the other fibers. Correspondingly, the spectral reflectivity of wool is

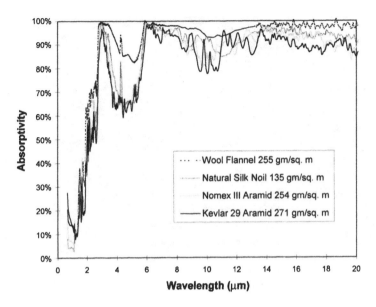

FIG. 6 Spectral absorptivity versus wavelength for polyamides at standard conditions.

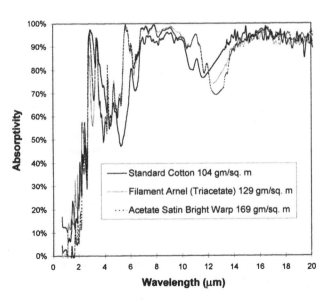

FIG. 7 Spectral absorptivity versus wavelength for cellulosics at standard conditions.

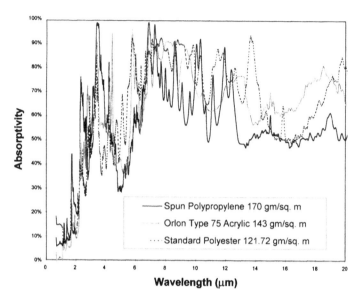

FIG. 8   Spectral absorptivity versus wavelength for hydrophobics at standard conditions.

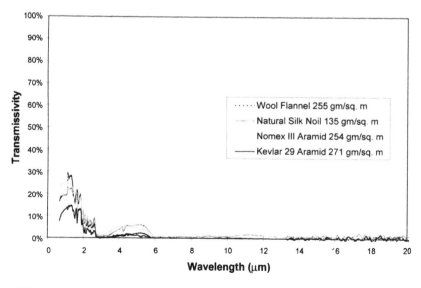

FIG. 9   Spectral transmissivity versus wavelength for polyamides at standard conditions.

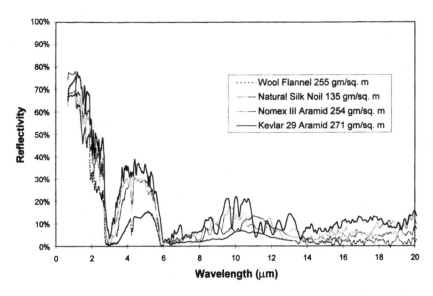

FIG. 10   Spectral reflectivity versus wavelength for polyamides at standard conditions.

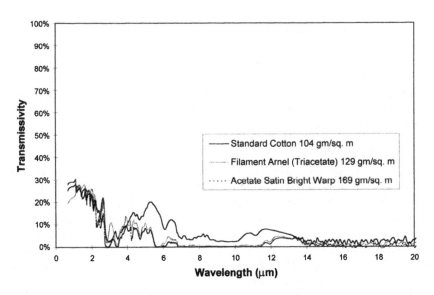

FIG. 11   Spectral transmissivity versus wavelength for cellulosics at standard conditions.

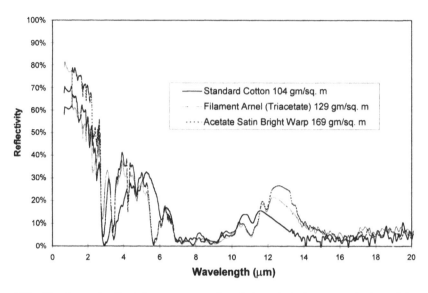

FIG. 12 Spectral reflectivity versus wavelength for cellulosics at standard conditions.

lower in this region. This might be explained by wool's higher regain at standard conditions; however, the spectral absorptivity of wool is also higher for dry fiber. The polar groups responsible for the high moisture regain of wool may be responsible for the higher absorptivity of dry wool.

The trends in the spectral absorptivity, spectral transmissivity, and spectral reflectivity plots (Figs. 8, 13, and 14) for the hydrophobic fibers are similar, but there are large local differences. The hydrophobic fibers absorb less, transmit more, and reflect more infrared radiation than the hydrophilic fibers in most regions of the spectrum.

3. Moisture Regain

The effects of moisture regain (ratio of weight of water to weight of dry fabric) on the spectral absorptivities, spectral transmissivities, and the spectral reflectivities of the hydrophilic fibers (wool and cotton) are less than on the hydrophobic fibers (polyester and polypropylene). This may be related to the polar groups in cotton and wool that are responsible for the high moisture regain in these fibers. The polar groups contribute to the spectral absorptivity of wool and cotton in a similar manner as water. Thus, the spectral characteristics of these fibers are less sensitive to the presence of water.

Moisture regain effects on the spectral absorptivities are illustrated in Figs. 15–18. The presence of moisture increases spectral absorptivities for all of the

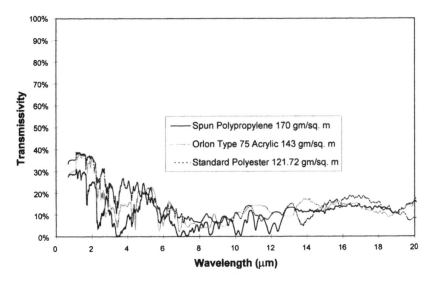

FIG. 13 Spectral transmissivity versus wavelength for hydrophobics at standard conditions.

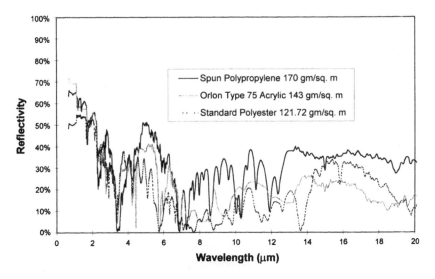

FIG. 14 Spectral reflectivity versus wavelength for hydrophobics at standard conditions.

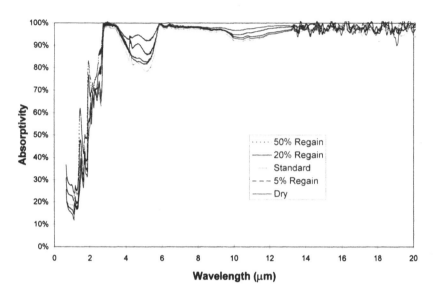

FIG. 15   Spectral absorptivity versus wavelength for wool 173 g/m² at various regains.

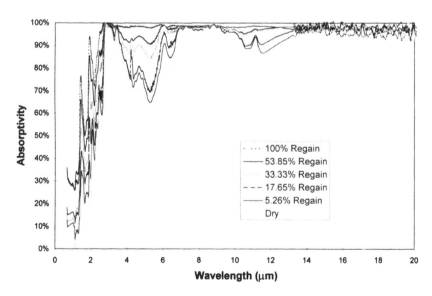

FIG. 16   Spectral absorptivity versus wavelength for cotton 193 g/m² at various regains.

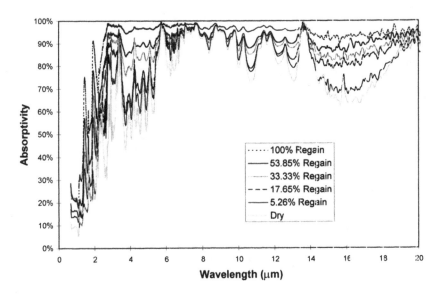

FIG. 17 Spectral absorptivity versus wavelength for polyester 174 g/m² at various regains.

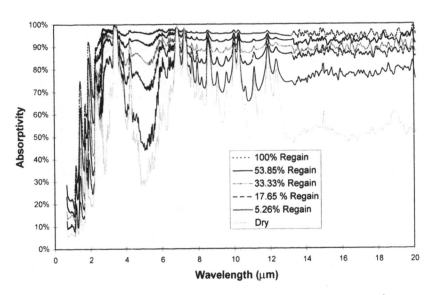

FIG. 18 Spectral absorptivity versus wavelength for polypropylene 170 g/m² at various regains.

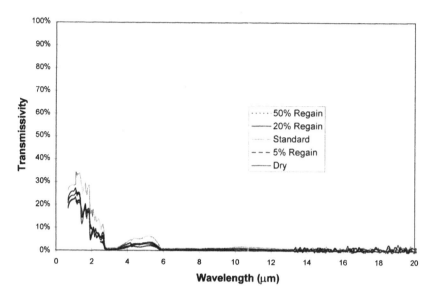

FIG. 19   Spectral transmissivity versus wavelength for wool 173 g/m² at various regains.

fabrics throughout the spectrum, but the spectral absorptivities in the NIR were much lower than in the rest of the spectrum. The effects of moisture on α_λ were larger for the hydrophobic fibers (polyester and polypropylene) than the hydrophilic fibers (wool and cotton) for wavelengths greater than 2.5 µm. The effects on cotton were larger than on wool. For wavelengths greater than 2.5 µm, the effects of moisture on the spectral absorptivities of wool were small.

At 100% moisture regain, the spectral absorptivity plots of all fabrics are nearly identical, and α_λ approaches 100% for wavelengths from 2.5 to 20 µm. The spectral absorptivities of the fabrics having a moisture regain of 30% are also large (typically greater than 90%) for wavelengths greater than 6 µm. At wavelengths between 2.5 and 6 µm, spectral absorptivities are high, but typically less than 90%.

Spectral transmissivities of wool, cotton, polyester, and polypropylene (Figs. 19–22) decrease with increasing moisture regain for wavelengths greater than 2.5 µm. Above 6 µm, the spectral transmissivity is almost zero at all regains for the heavier fabrics. However, in this region, polyester and polypropylene transmit some radiation when the fabrics are dry or nearly dry. Moisture does not diminish spectral transmissivity in the NIR as it does in the other regions. For wavelengths below 2.5 µm, spectral transmissivities for all of the fibers with moisture regain of 100% are typically 30–35%.

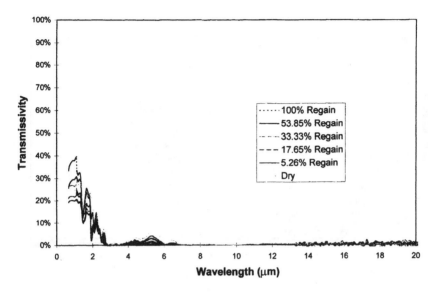

FIG. 20 Spectral transmissivity versus wavelength for cotton 193 g/m² at various regains.

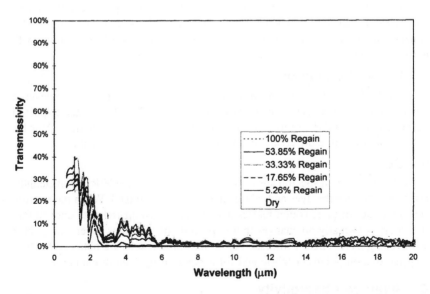

FIG. 21 Spectral transmissivity versus wavelength for polyester 174 g/m² at various regains.

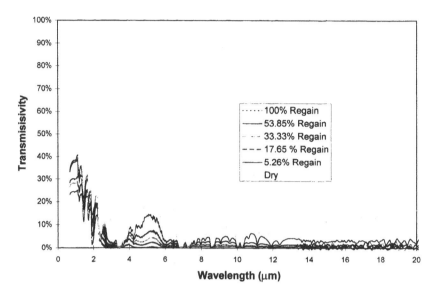

FIG. 22 Spectral transmissivity versus wavelength for polypropylene 170 g/m² at various regains.

Spectral reflectivities of wool, cotton, polyester, and polypropylene (Figs. 23–26) decrease with increasing moisture regain throughout the spectrum, but are nonzero in the NIR. Values in the NIR are as high as 30–40% for fabrics with moisture regain of 100%.

4. Fabric Construction

Most of the fabrics that have been studied have been woven fabrics; however, a few knit fabrics have been compared to woven fabrics. Knit fabrics have spectral characteristics similar to those of comparable weight woven fabrics. Data for other structures quite different from woven fabrics are not available.

5. Dye

The effects of dye on spectral absorptivities are limited to the NIR at wavelengths less than 1 μm for the dyes and fabrics that have been studied. When the spectral characteristics of three cotton fabrics dyed black are compared with undyed fabrics, there is virtually no changes in the spectral absorptivities except in the NIR at wavelengths less than 1 um (see Fig. 27). Similar results are obtained for blue disperse dye on polyester fabric and for a range of dyes on nylon carpet.

C. Average Absorptivity

The integrated average absorptivities presented in this subsection are for woven fabrics and blackbody emitters. Results for a few knit fabrics indicate that knit

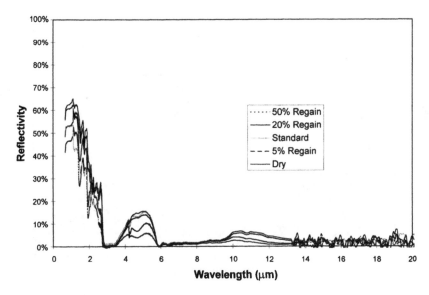

FIG. 23 Spectral reflectivity versus wavelength for wool 173 g/m² at various regains.

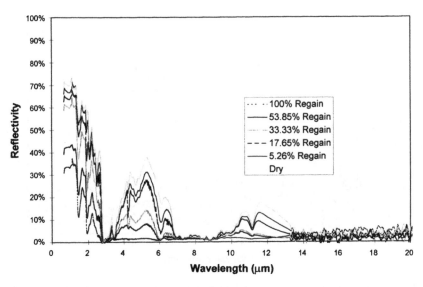

FIG. 24 Spectral reflectivity versus wavelength for cotton 193 g/m² at various regains.

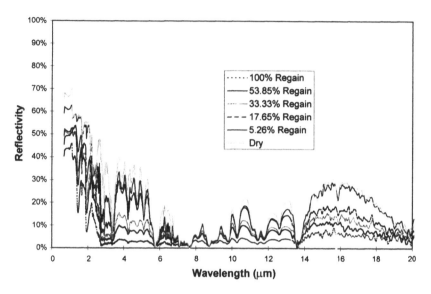

FIG. 25 Spectral reflectivity versus wavelength for polyester 174 g/m² at various regains.

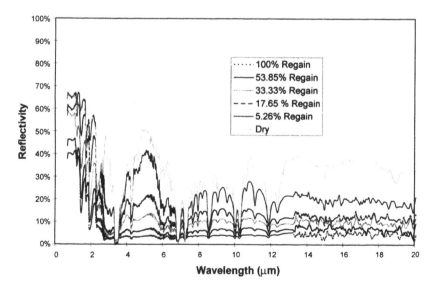

FIG. 26 Spectral reflectivity versus wavelength for polypropylene 170 g/m² at various regains.

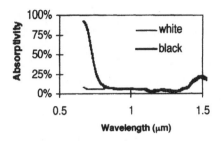

FIG. 27 Spectral absorptivity of dyed and undyed cotton fabric.

fabrics behave similarly to woven fabrics of comparable weight. Limited data [3,4] indicate that integrated average absorptivities of structures such as pile fabric may be significantly different from those of woven fabrics.

1. Emitter Temperature

The effect of blackbody emitter temperature on integrated average absorptivity, $\bar{\alpha}$, is illustrated in Fig. 28. Integrated average absorptivity decreases with increasing emitter temperature. Radiation emitted by low-temperature (long-wavelength) blackbody emitters is spread throughout the spectrum at wavelengths

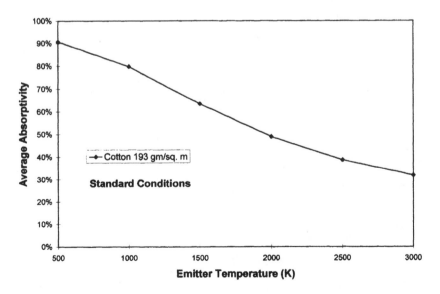

FIG. 28 Typical variation of average absorptivity with blackbody emitter temperature.

greater than approximately 2 μm where fabric spectral absorptivities are typically high. Thus, integrated average fabric absorptivities are high for the low-temperature emitters. On the other hand, the radiation emitted by the high-temperature emitters is concentrated in the shorter-wavelength region of the spectrum where spectral absorptivities of the fabrics are low. Consequently, $\bar{\alpha}$ is lower for short-wavelength emitters.

2. Fabric Weight (Areal Density)

The effect of fabric weight (areal density) on integrated average absorptivity is illustrated in Fig. 29, which shows the variation in $\bar{\alpha}$, for two blackbody emitter temperatures. As would be expected, the lower weight fabrics have significantly lower integrated average absorptivities. As fabric weight is increased, $\bar{\alpha}$ increases rapidly at first, then a threshold (approximately 100 g/m^2) is reached where further increase in fabric weight has little effect.

3. Fiber Type

The effect of fiber type on $\bar{\alpha}$ depends greatly on moisture regain as will be discussed in the next section. For a blackbody emitter temperature of 1500 K, integrated average absorptivity is plotted against fabric weight for standard conditions in Fig. 30. The average absorptivities of the hydrophilic fibers (cellulosics and polyamides) are similar, and they are higher than for the hydrophobic fibers (polyester, polypropylene, and acrylic). This is associated with the lower regain

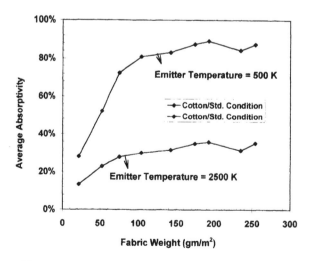

FIG. 29 Average absorptivity versus fabric weight for 500 K and 2500 K emitter temperatures.

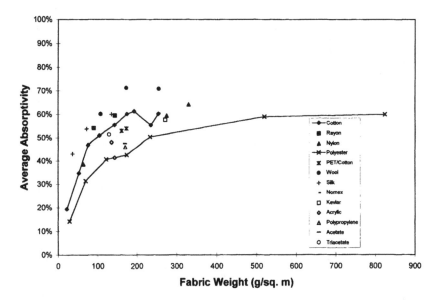

FIG. 30   Fabric weight versus average absorptivity for emitter temperature = 1500 K.

of the hydrophobic fibers at standard conditions. Integrated average absorptivities of cotton/polyester blends are between those of cotton and polyester fabrics, but are closer to those of cotton fabric.

4.   Moisture Regain

The effect of moisture regain on $\bar{\alpha}$ is shown in Figs. 31 and 32 for blackbody emitter temperatures of 500 K and 2500 K, respectively. Integrated average absorptivity increases with increasing moisture regain for all of the fibers, but the effects are largest for the hydrophobic fibers. As moisture regain increases from 0% to 25%, $\bar{\alpha}$ rapidly increases for polypropylene, but increases only slightly for wool. Above 25% moisture regain, the average absorptivities of all fabrics begin to level off, especially for blackbody emitter temperatures of 500 K. The average absorptivities for both hydrophobic and hydrophilic fibers are similar in the region above 25%. The fiber least affected by moisture is wool, which had the highest regain at standard conditions.

For moisture regain of 100% and fabric weight above 100 g/m², $\bar{\alpha}$ for a blackbody emitter temperature of 500 K is close to 100%. Under the same conditions, integrated average absorptivities for emitter temperatures of 1500 K and 2500 K are approximately 80% and 60%, respectively. A significant amount of the output of the high-temperature blackbody emitters is in the NIR, where the spectral absorptivities of the fabrics are low even when the fabrics are wet.

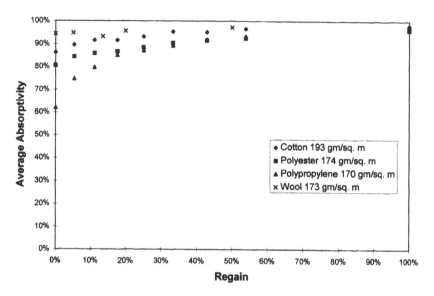

FIG. 31 Regain versus average absorptivity for 500 K emitter temperature.

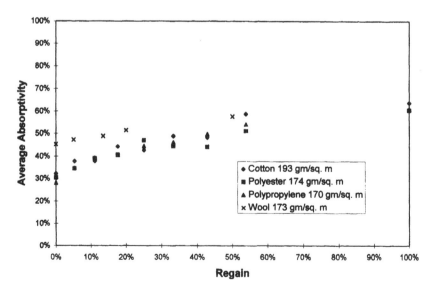

FIG. 32 Regain versus average absorptivity for 2500 K emitter temperature.

D. Overall Radiant Efficiency

Two factors are important in evaluating the overall radiant efficiency of IR emitters. One is radiant efficiency of the IR emitter (how efficiently the emitter converts input energy into IR energy). The other is how well emissions of the IR emitter is absorbed by the fabric. Short-wavelength (high-temperature) blackbody emitters are better at converting electrical energy into IR radiation, but high-wavelength (low-temperature) blackbody emitters have radiation spread throughout the spectrum where fabric spectral absorptivities are higher. In calculating overall radiant efficiency (ξ), both of these factors are taken into account.

The effect of emitter temperature on overall radiant efficiency, ξ, for four types of dry fabrics is shown in Fig. 33. For all the fabrics, ξ peaks at a temperature of about 1500 K but does not change much between 1000 K and 2000 K. For dry fibers, the overall radiant efficiency is higher for the hydrophilic fibers. At the high moisture regain, ξ is almost identical for all of the fibers.

The typical variation of overall radiant efficiency with emitter temperature for several moisture regains is shown in Figs. 34 and 35. As moisture regain is increased, ξ increase for both the hydrophilic and hydrophobic fibers, but the increase is greater for the hydrophobic fibers. Also, the effects are larger for emitters temperature at and above 1500 K. At the high moisture regains and blackbody emitter temperature above 1500 K, ξ varies little with emitter temperature.

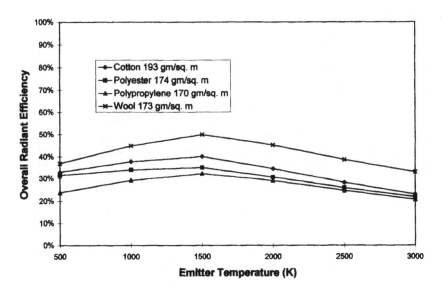

FIG. 33 Overall radiant efficiency versus emitter temperature for dry fabrics.

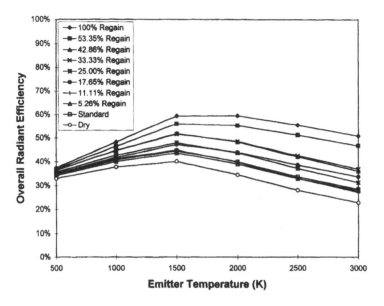

FIG. 34 Overall radiant efficiency versus emitter temperature for cotton 193 g/m².

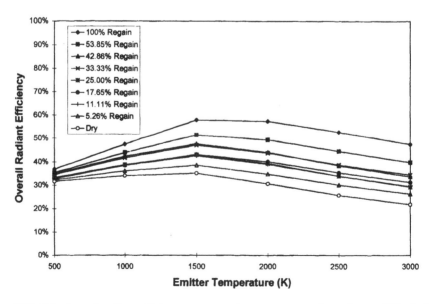

FIG. 35 Overall radiant efficiency versus emitter temperature for polyester 174 g/m².

REFERENCES

1. W. W. Carr, D. S. Sarma, M. R. Johnson, B. T. Do, and V. A. Williamson. Text. Res. J. (1997).
2. E. McFarland, Master's thesis, Georgia Tech (1997).
3. A. Alkidas, E. R. Champion, W. E. Giddens, R. W. Hess, B. Kumar, G. A. A. Naveda, P. D. Durbetaki, P. T. Williams, and W. Wulff, Second Final Report Georgia Institute of Technology, December 31, 1972 (NTIS: COM-73-10956).
4. S. Backer, G. C. Tesoro, T. Y. Toong, and N. A. Moussa, in *Textile Fabric Flammability*, MIT Press, Cambridge, MA, 1976, pp. 22–106.
5. A. D. Broadbent, B. Cote, T. Fecteau, P. Khatibi-Sarabi, and N. Therien. Text. Res. J. *64*:123 (1994).
6. L. Campagna, B. Chotard, N. Christen, L. Godin, A. Houle, and R. Nantel, in *AATCC Book of Papers*, 1988, Vol. 46, pp. 47–56.
7. C. Langlois and R. Maisonneuve, AATCC International Conference and Exhibition, 1988, pp. 85–94.
8. W. H. Rees and L. W. Ogden. Text. Inst. J. 13 (1946).
9. W. Wulff, N. Zuber, A. Alkidas, and R. W. Hess. Combustion Sci. Technol. *6*:321 (1973).
10. F. P. Incropera, and D. P. Dewitt, in *Fundamentals of Heat and Mass Transfer*, 3rd ed., Wiley, New York, 1990, Chap. 12.
11. Electric infrared process heating: State-of-art assessment, EPRI Report EM-4571, (March 1987).
12. M. Orfeuil, in *Electric Process Heating*, Batelle Press, Columbus, OH, 1987, Chap. 5 (EPRI EM-5105-SR).
13. *Technology Guidebook for Electric Infrared Process Heating*, Electric Power Research Institute, Center for Materials Fabrication, 1993.
14. M. L. Toison, in *Infrared and Its Thermal Applications*, N. V. Phillips, Eindhoven, 1964, Chap. 5.
15. Fastoria Industries, Inc., *Electric Infrared for Industrial and Process Heating Applications*, Bulletin 50-580-87, Fastoria Industries, Inc., 1987.
16. M. J. Lampinen, K. T. Ojala, and E. Koski. Drying Technol. *9*:973 (1991).
17. K. T. Ojala, E. Koski, and M. J. Lampinen. Appl. Opt. *31*:4589 (1992).
18. R. R. Willey. SPIE Infrared Technol. Applic. *590*:248 (1985).

5

Electrochemical Sensors for the Control of the Concentration of Bleaching Agent to Optimize the Quality of Bleached and Dyed Textile Products

PHILIPPE WESTBROEK and EDUARD TEMMERMAN Department of Analytical Chemistry, University of Gent, Gent, Belgium

PAUL KIEKENS Department of Textiles, University of Gent, Gent, Belgium

I. INTRODUCTION

The term "cleaning of textiles" is understood as the series of treatments at the textile material to obtain a white, uncontaminated product. During these treatments, many contaminating products are removed from the textile product. The treatments included in the term *cleaning* are as follows:

Unstrengthening of the textile product to remove amylum, poly(vinyl alcohol), acrylates or carboxymethyl cellulose which were affixed before weaving to strengthen the fiber against deformation and breakage [1]

Washing and cooking [2] to remove natural fats, oils, pectines, hemicelluloses, proteins, mineral compounds, and sugars

Oxidative [3] or reductive [3] bleaching to remove all the natural coloring pigments which cannot be removed by the preceeding treatments

It is clear that all these processes have an influence on the quality and uniformity of the treated textile product but the last mentioned process (bleaching) is the most important because it directly fixes the quality (degree of whiteness [4] and degree of desizing) of the textile product. This quality is strongly dependent on the conditions used in the process, such as concentration and type of bleaching agent, pH, temperature, and additives used [5]. Therefore, it is clear why this chapter is dedicated to the bleaching of textiles and how one can achieve the best quality for a textile product.

The goal of a bleaching process is as follows:

To obtain a textile product that can take up water, dyes, and dressing agents in an equal and reproducible way

A high, permanent, and reproducible degree of whiteness, without degradation of the textile structure itself, to guarantee the color fastness of the used dyes during the following dyeing process.

II. BLEACHING AGENTS

Up to 20 years ago, NaOCl and $NaClO_2$ were the most used bleaching agents because of their excellent bleaching properties [6]. Due to ecological imparatives, their major role is gradually taken over by hydrogen peroxide [7] because its reaction products after bleaching are water and oxygen [8] instead of the Cl^--containing compounds from NaOCl and $NaClO_2$.

Therefore, only minor attention will be given to these products, and a more detailed description is given about the role and importance of hydrogen peroxide in bleaching. Brief attention will also be given to bleaching processes based on enzymes [9], which are not common in bleaching of textiles but find their industrial application in the bleaching of paper pulp [9].

A. Sodium Hypochlorite

The synthesis of sodium hypochlorite is based on pumping gaseous chlorine through a sodium hydroxide solution [10]:

$$2NaOH + Cl_2 \leftrightarrow NaOCl + NaCl + H_2O \tag{1}$$

followed by isolation of NaOCl.

By dissolving the solid sodium hypochlorite into the aqueous solution of the bleaching bath, it hydrolyzes as illustrated in reactions (2) and (3):

$$NaOCl + H_2O \leftrightarrow HOCl + NaOH \tag{2}$$

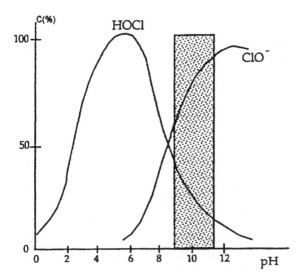

FIG. 1 Dissociation diagram of a HOCl solution as a function of pH.

$$HOCl \leftrightarrow HCl + [O] \tag{3}$$

The atomic oxygen [O] (also called active oxygen) is the compound that is directly involved in the bleaching reaction and possesses extremely high oxidative properties. As can be seen from Fig. 1, the highest concentration of HOCl (and [O]) is obtained in a pH range from 4 to 6, which results in strong bleaching effects (a high degree of whiteness) but potentially also in damage to the textile structure caused by the unselective behavior of sodium hypochlorite. Therefore, it is very important to obtain an optimal concentration of the active oxygen [O] to get satisfactory bleaching effects without degradation of the textile structure itself. This can be achieved by keeping the pH in the range of 9–11.5, as can be seen from Fig. 1.

For bast fibers, to the contrary, a pH lower than 5 is used because the colored impurities present in the fiber are chlorinated by the reaction with sodium hypochlorite and show less degradation of the fiber compared to cotton. These chlorinated compounds can easily be removed in alkaline solutions because of their good solubility in these solutions. Therefore, the bleaching of bast fibers is carried out in slightly acidic medium followed by an alkaline washing process.

Almost all bleaching processes based on sodium hypochlorite are carried out at room temperature. It is clear that the bleaching effects would increase with increasing temperature, but a stronger increase of the textile damage is also observed.

B. Sodium Chlorite

The synthesis of sodium chlorite is based on a treatment of ClO_2 in strongly alkaline solution [10] by using sodium hydroxide,

$$2ClO_2 + 2NaOH \leftrightarrow NaClO_2 + NaClO_3 + H_2O \tag{4}$$

or by using sodium peroxide [10],

$$2ClO_2 + Na_2O_2 \leftrightarrow 2NaClO_2 + O_2 \tag{5}$$

In preparing the bleaching solution, sodium chlorite hydrolyzes with water but not so effectively as sodium hypochlorite because hydrogen chlorite is a stronger acid ($K_a = 1.1 \times 10^{-2}$ [11]) than hydrogen hypochlorite ($K_a = 3.2 \times 10^{-8}$ [11]). To the contrary, the active oxygen formed from sodium chlorite is very selective for the colored impurities, which means that a high concentration can be used without degradation or damage of the textile structure itself. The optimal pH for obtaining high concentrations of $HClO_2$ which decomposes into active oxygen [reaction (6)] is between 2 and 3.5:

$$HClO_2 \leftrightarrow HCl + 2[O] \tag{6}$$

Nevertheless, side reactions of $HClO_2$ take place at pH < 3 [reactions (7)–(9)].

$$2HClO_2 \leftrightarrow HClO_3 + HClO \tag{7}$$
$$HClO_2 + HClO \leftrightarrow HClO_3 + HCl \tag{8}$$
$$HClO_3 + HClO_2 \leftrightarrow 2\ ClO_2 + H_2O \tag{9}$$

As a result, the optimal pH value for bleaching moves from pH = 3 to pH = 4.5.

Bleaching processes based on sodium chlorite cannot be done in common bleaching machines because of a high corrosion rate. Only bleaching baths constructed from glass, ceramic materials, titanium alloys, or stainless steel with a high degree of molybdenum are resistant to solutions containing sodium chlorite.

C. Hydrogen Peroxide

1. Alkaline Bleaching Processes

Hydrogen peroxide can be synthesized in different ways. The most used method is the auto-oxidation procedure [10] employing 2-ethyl antra quinone (Fig. 2). Other possibilities are [10]

$$BaO_2 + H_2SO_4 \leftrightarrow H_2O_2 + BaSO_4(s) \tag{10}$$

or [10]

$$2(NH_4)HSO_4 \leftrightarrow (NH_4)_2S_2O_8 + H_2(electrolysis) \tag{11}$$

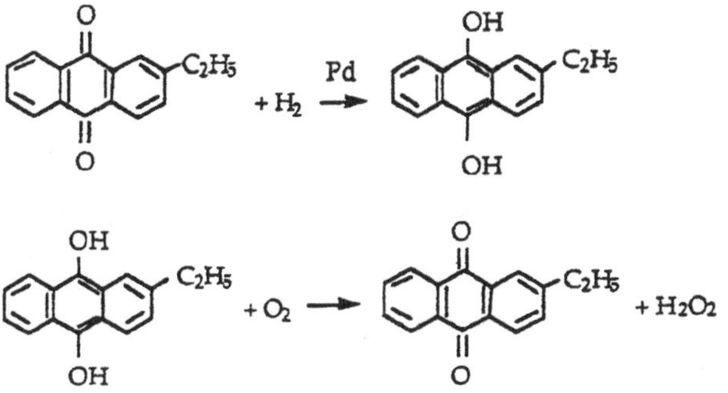

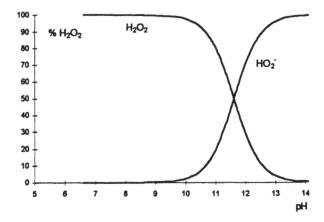

FIG. 2 Synthesis of hydrogen peroxide by the auto-oxidation procedure.

$$(NH_4)_2S_2O_8 + 2\,H_2O \leftrightarrow 2(NH_4)HSO_4 + H_2O_2 \tag{12}$$

As already mentioned in Section I, hydrogen peroxide has become the most important bleaching agent over the last 20 years because of its ecological benefits. Hydrogen peroxide itself (H_2O_2) possesses only weak bleaching properties [12] compared to its conjugated base HO_2^-. Hydrogen peroxide belongs to the class of very weak acids [12] in aqueous solution. This is immediately evident when one compares its acidity constant ($K_a = 2.5 \times 10^{-12}$ [13]) with that of water itself ($K = 1 \times 10^{-14}$ [14]). Therefore, a strongly alkaline medium is needed to obtain important concentrations of the conjugated base [reaction (13) and Fig. 3]:

FIG. 3 Dissociation diagram of hydrogen peroxide in aqueous solution as a function of pH.

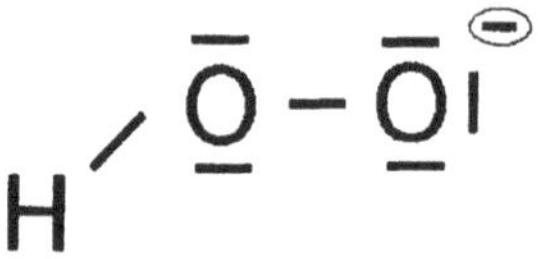

FIG. 4 Structure and electron configuration of hydrogen peroxide in strongly alkaline solutions.

$$H_2O_2 \leftrightarrow HO_2^- + H^+ \tag{13}$$

Therefore, in practice, bleaching processes based on hydrogen peroxide are carried out in a pH range from 10 to 14 [15]. To obtain this pH, sodium hydroxide is commonly used [15].

From specific research [16], among other experiments with radical catchers, the presence of an oxygen radical anion $O_2^{\cdot-}$ was demonstrated which is directly involved in the bleaching reaction and is formed out of the conjugated base HO_2^-. The radical anion is easily formed, whereas HO_2^- is unstable due to the localized charge [10] at the end oxygen (see Fig. 4).

The mechanism of the reactions of the radical anion has not been unraveled completely and is relatively complicated because of the reactivity of the $O_2^{\cdot-}$ [17]. It first reacts with colored impurities in the textile structure [18]; second, decomposition reactions occur [18]; and, finally, it reacts with the textile structure itself because of low selectivity [19].

The bleaching reaction itself is an oxidation [18] of the chromophore groups present in the textile structure by $O_2^{\cdot-}$ which is formed out of hydrogen peroxide. Due to this oxidation, the wavelength of absorption of these groups moves out of the UV-vis region which can be translated into a change from a colored to an uncolored product [20].

Similar to sodium hypochlorite, the active compound responsible for the bleaching effects possesses only a small selectivity toward the colored impurities, which means that the excess also attacks the textile structure itself. At a pH value of 11, there is already sufficient active oxygen present [17] to cause damage to textile goods. On the other hand, at a pH value of 10.9, there is not enough hydrogen peroxide deprotonated [reaction (13) and Fig. 3] and therefore the bleaching effects are very low.

To avoid this problem, the formation of $O_2^{\cdot-}$ is stabilized by forming complexes of HO_2^- with magnesium ions [21], which are kept soluble [a high pH gives the slightly soluble $Mg(OH)_2$] by using sodium silicate [21]. Due to the formation of this complex, the HO_2^- has stabilized because the localized charge at the end oxygen atom (Fig. 4) is now more spread over the complexed compound. A drawback of the use of sodium silicate is the polymerization to insoluble

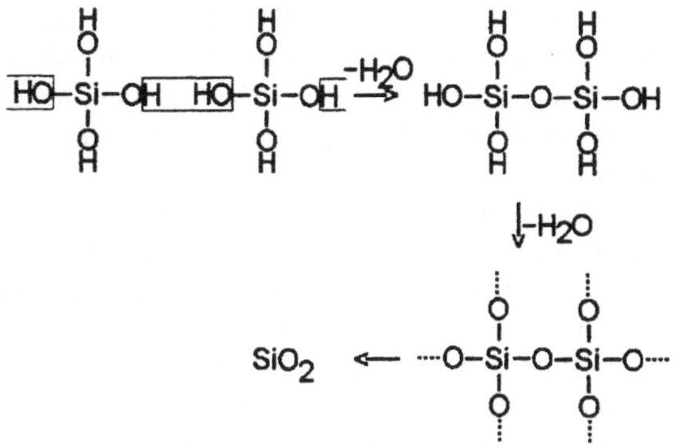

FIG. 5 Polymerization mechanism of sodium silicate at high temperature and/or high concentration.

polymers (Fig. 5), which causes precipitates that are difficult to remove at the textile goods and the bleaching machine. Despite this drawback, sodium silicate is still largely used in bleaching processes, not only for its important role in the stabilization of HO_2^- [21] but also for its inactivation of metal ions [21] (and their oxides and hydroxides) present in the bleaching solution and in the textile structure. These ions are good catalysts for the decomposition of hydrogen peroxide into oxygen [17]:

$$2H_2O_2 \leftrightarrow O_2 + 2H_2O \tag{14}$$

The reaction intermediates of this decomposition (such as OH˙ radicals) have extremely strong oxidizing properties [19] and cause damage and degradation of the textile structure. In the presence of sodium silicate, these metal ions and their compounds lose their property to catalyze the decomposition of hydrogen peroxide.

It is observed that the polymerization of sodium silicate becomes important if the temperature and/or the used concentration of sodium silicate is relatively high [21]. To circumvent this effect, a new type of so-called stabilizers is developed which are based on the EDTA structure (Fig. 6) [6]. Their ability to stabilize

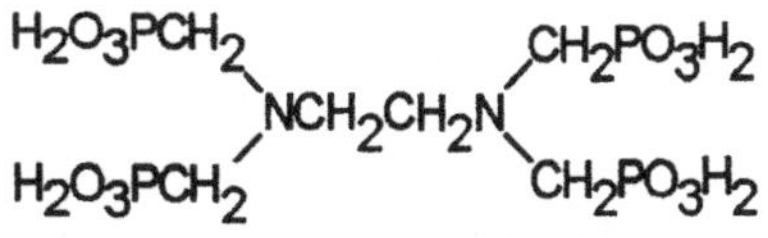

FIG. 6 Structure of stabilizers based on EDTA.

the formation of active oxygen is similar to that of sodium silicate, but not so explicit [6]. Therefore, these new stabilizers are used in combination with sodium silicate in order to obtain an optimal stabilization effect of the bleaching liquor without polymerization of sodium silicate, even at high temperatures, because of the use of a smaller concentration.

In practice, two main types of bleaching process are carried out, namely the dry–wet and the wet–wet bleaching process. This nomenclature defined the condition of the textile goods. In a dry–wet process, the textile fabric enters the bath in a dry state and leaves it wet; in the second process, the wetted textile good enters the bath. The wet–wet process has the advantage that the energy barrier between bleaching solution and textile good is smaller [22,23], but for both processes, surfactants are added to the bleaching solution to minimize this energy barrier. A drawback of these surfactants is the formation of foam. For minimizing this problem, antifoam products [24] are added to the bleaching solution.

With a bleaching solution containing hydrogen peroxide, sodium hydroxide, magnesium ions, sodium silicate, surfactants, and antifoam products, the best bleaching quality can be obtained in the shortest possible time if the concentration of hydrogen peroxide is kept constant. A too small concentration gives insufficient bleaching effects (a low degree of whiteness) and a too high concentration causes degradation of the textile structure (a good degree of whiteness but a too high degree of desizing). Therefore, it is needed to measure and to control the hydrogen peroxide concentration and, if possible, in a continuous and in-line way.

Despite many methods [25] that have been developed and described in literature, the titration by hand of bath probes with potassium permanganate [26] is still the most used method in the industry (Fig. 7). This method has a good precision and accuracy if the titration is carried out correctly. Often, this is not the case; possible causes are the following:

$$2(MnO_4^- + 8H^+ + 5e^- \rightleftharpoons Mn^{2+} + 4H_2O)$$
$$5(H_2O_2 \rightleftharpoons O_2 + 2H^+ + 2e^-)$$
$$\overline{2MnO_4^- + 6H^+ + 5H_2O_2 \rightleftharpoons 2Mn^{2+} + 8H_2O + 5O_2}$$

$$c_{H_2O_2} = \frac{5c_{MnO_4^-} V_{MnO_4^-}}{2V_{H_2O_2}}$$

FIG. 7   Reactions involved in the determination of the hydrogen peroxide concentration by titration of a bath sample with potassium permanganate.

Incorrect procedures for taking the bath probes.

The place where the probe is taken (in the bath, in the mixing tank, in a bypass) is very important, as concentration gradients are normal.

Potassium permanganate decomposes slowly [27], which may cause erroneous results.

Incorrect calculation of the hydrogen peroxide concentration from the consumed volume of potassium permanganate.

Frequency of titrations too small.

The titration is based on a visual detection of the end point. This may lead to a different result between titrations done by different workers.

Trials to use automatic titrators failed because of irreproducible results [28] and the relatively high price of an automatic titrator [29]. Other techniques such as potentiometry [30], calorimetry [29], and conductometric methods were not accurate or precise enough, or the price was too high, as for the colorimetric technique. A sensor system has been developed at the University of Gent, Belgium which can measure and control the hydrogen peroxide concentration during bleaching in a continuous and in-line way without sample-taking. The sensor reacts immediately at a change of the hydrogen peroxide concentration, which gives the opportunity to control the concentration and to obtain the best possible bleaching effects (the highest degree of whiteness) without degradation of the textile structure (the lowest degree of desizing).

In an industrial environment, the hydrogen peroxide sensor is positioned in a bypass of the bleaching bath. The bypass is provided with five holes (Fig. 8) for implementation of the electrodes, namely a combined pH sensor [32], reference electrode (Ag/AgCl/Cl$^-$) [32], and counterelectrode [32] (Pt rod), a PT100

FIG. 8 Scheme of the bypass part in which the electrodes are positioned for measurement of the hydrogen peroxide concentration in an industrial environment.

probe, and the working electrode. This sensor electrode is a carbon rod embedded in epoxy resin and pretreated by a special procedure [33].

At an electron conductor, which acts as a working electrode, hydrogen peroxide can be oxidized or reduced [17]. With a potentiostat, a constant potential difference can be applied between the working and the reference electrodes and the current can be measured between the working electrode and counterelectrode. It was found that most of the materials commonly used as working electrodes cannot be used in bleaching baths, because the linearity between voltametric or amperometric signal and hydrogen peroxide concentration is limited to 0.085 g/L due to ohmic drop effects [33]. Concentrations used in bleaching baths can vary from 1 up to 70 g/L. From preliminary experiments, it was clear that a carbon electrode gave promising results [33] concerning the measurement of hydrogen peroxide concentrations up to sufficiently high values. To achieve this, use was made of a special oxidation reaction of hydrogen peroxide at a carbon electrode pretreated in a specific way (Fig. 9). The mechanism of this reaction yields a low current density, which minimizes the ohmic drop effect. The linearity between signal and hydrogen peroxide concentration is guaranteed up to concentrations higher than 70 g/L due to this lower current density. There is one drawback to take into account: The special reaction occurs only if pH > 10.5. For use in

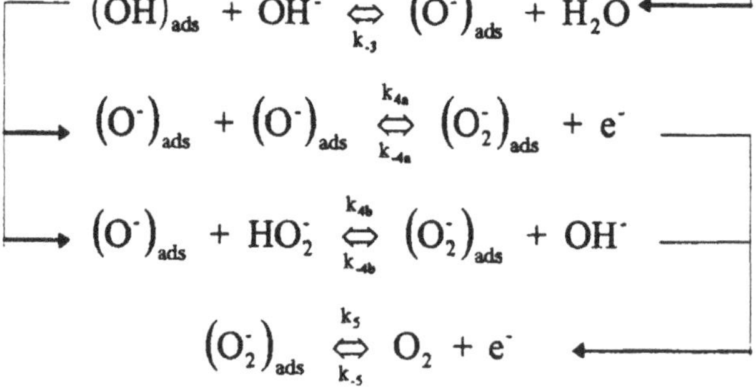

$$HO_2^- \overset{k_1}{\underset{k_{-1}}{\Leftrightarrow}} \left(HO_2^-\right)_{ads}$$

$$\left(HO_2^-\right)_{ads} \overset{k_2}{\underset{k_{-2}}{\Leftrightarrow}} \left(O^-\right)_{ads} + \left(OH\right)_{ads}$$

$$\left(OH\right)_{ads} + OH^- \overset{k_3}{\underset{k_{-3}}{\Leftrightarrow}} \left(O^-\right)_{ads} + H_2O$$

$$\left(O^-\right)_{ads} + \left(O^-\right)_{ads} \overset{k_{4a}}{\underset{k_{-4a}}{\Leftrightarrow}} \left(O_2^-\right)_{ads} + e^-$$

$$\left(O^-\right)_{ads} + HO_2^- \overset{k_{4b}}{\underset{k_{-4b}}{\Leftrightarrow}} \left(O_2^-\right)_{ads} + OH^-$$

$$\left(O_2^-\right)_{ads} \overset{k_5}{\underset{k_{-5}}{\Leftrightarrow}} O_2 + e^-$$

FIG. 9 Reaction mechanism of the oxidation of hydrogen peroxide in alkaline solutions at a properly pretreated carbon electrode.

textile bleaching baths, this is not a problem because these processes mostly take place at pH > 10.5.

The current measured at the carbon electrode after pretreatment is proportional to the hydrogen peroxide concentration but is also dependent on pH and temperature. This means that every datum point has to be compensated for pH and temperature differences between the actual values and the values of the same parameters at the calibration point. This also means that pH and temperature have to be measured continuously and this is why a temperature sensor and pH sensor are implemented in the bypass. Compensation of the experimental signal for pH differences can be done in bleaching processes with the following compensation formula [34]:

$$\log I_{pH} = 0.3001(pH_{cal} - pH_m) + \log I_m \qquad (15)$$

I_m and I_{pH} are respectively the experimental voltametric current and the signal compensated for pH, respectively, whereas pH_{cal} and pH_m are the pH values of the calibration point and the experimental point, respectively. For measuring concentrations of hydrogen peroxide lower than 15 g/L, the compensation formula remains the same except for the numerical factor 0.3001.

The relation between sensor output signal and temperature is additionally complicated because the pH value is also temperature dependent. Before obtaining information about temperature dependence, the signals must first be compensated for pH. As for the pH compensation formula (15), an expression was deduced for the compensation of the signal for temperature changes [34]:

$$\log I_{pH, T} = 0.9002(\log T_{cal} - \log T_m) + \log I_{pH} \qquad (16)$$

where $I_{pH, T}$ is the signal compensated for bath pH and temperature and T_{cal} and T_m are the temperature values at the calibration point and the experimental point, respectively. The sensor system is implemented as a bypass at the bleaching machine (Fig. 10). This position is choosen on the one hand because of laminar-flow profiles in the bypass and at the surfaces of the electrodes and on the other hand because of the optimal position to control additions of hydrogen peroxide and sodium hydroxide in the mixing tank during the running process.

2. Acid Bleaching Processes

As already described in the previous subsection, hydrogen peroxide itself has only weak bleaching properties. As a result, direct bleaching with hydrogen peroxide can only be done in alkaline medium. However, hydrogen peroxide is also used in bleaching processes in acidic medium (e.g., for the bleaching of wool), but in an indirect way. Hydrogen peroxide does not act as a bleaching agent itself but as an oxidizing substance for acetic acid anhydride [35] from which acetic acid and peracetic acid are formed [reaction (17)] [35]. The formed acetic acid

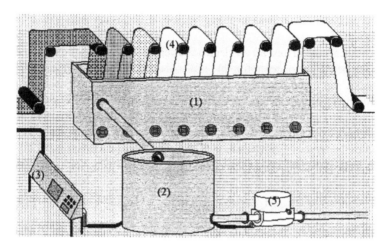

FIG. 10 Scheme of a bleaching setup and position of the hydrogen peroxide sensor system: (1) bleaching bath, (2) mixing tank, (3) control panels, (4) textile fabric, and (5) sensor system.

is further oxidized to peracetic acid by hydrogen peroxide [reaction (18)] [35]. The formed peracetic acid decomposes spontaneously to the acetic acid radical and a hydroxyl radical [reaction (19)] [35] which posesses strong oxidative properties.

$$CH_3C(=O)-O-(O=)CCH_3 + H_2O_2$$
$$\rightarrow CH_3C(=O)OOH + CH_3C(=O)OH \quad (17)$$
$$CH_3C(=O)OH + H_2O_2 \rightarrow CH_3C(=O)OOH + H_2O \quad (18)$$
$$CH_3C(=O)OOH \rightarrow CH_3COO^{\cdot} + {}^{\cdot}OH \quad (19)$$

Two main reasons for using the acidic bleaching process are the following:

This type of bleaching is very useful for textiles which are not stable in alkaline medium.

The active component in the bleaching process is very selective for the colored impurities, which means that even at high concentrations of bleaching agent, damage or degradation of the textile structure is not observed.

Despite these advantages, some precautions must be taken when using an acidic bleaching process based on acetic acid anhydride and hydrogen peroxide:

Peracetic acid is relatively volatile and irritates the mucous membranes, which means that a closed bleaching tank is needed.

The above-mentioned fact makes it difficult to use continuous and/or semicontinuous bleaching processes.

When the process is not properly controlled, the formation of the explosive product diacetylperoxide is possible.

The process is a lot more expensive compared with alkaline bleaching.

The hydrogen peroxide sensor developed at the University of Gent and described in the previous subsection cannot be used in this type of bleaching process because the sensor system only performs well in the pH range from 10.5 up to 14 [36]. Fortunately, a solution for this problem has been found by using a FIA system (Fig. 11). FIA stands for Flow, Injection, and Analysis. A constant flow of bleaching liquor is pumped (with a peristaltic pump) toward a mixing chamber (Flow) and mixed with a constant flow of a sodium hydroxide solution (Injection). This mixing occurs in a mixing chamber which, on the one hand, levels up the pH value and, on the other hand, dilutes the original bleaching liquor. By knowing the flow rate of the bleaching liquor and the sodium hydroxide stream, the degree of dilution can be calculated. After mixing, the resulting solution passes through a small chamber where the electrodes are positioned to measure the hydrogen peroxide concentration in the strongly alkaline solution. Knowing the degree of dilution, the original concentration of hydrogen peroxide in the bleaching bath can be obtained by calculation. The solution used for analysis has a high pH value and this cannot be recycled into the bleaching solution. To diminish this waste, the system was miniaturized to the level that approximately 1 L of waste solution is produced per hour.

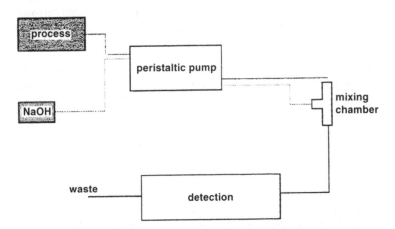

FIG. 11   Scheme of the modified sensor system using the FIA principle for measurement of the hydrogen peroxide concentration in processes where the pH is lower than 10.5.

D. Bleaching with Enzymes

Bleaching with enzymes is not used on a large scale for several reasons:

It is an expensive process.

It can only be used as a discontinuous process because the in situ obtained concentration of hydrogen peroxide is relatively low, which also means long interaction times and low productivity.

Enzymes are only active in a small pH range, which means that the pH of the process has to be controlled.

Because the enzymes are not inexpensive, they have to recuperated. A system to recuperate them from the used bleaching solution is an extra cost.

The role of the enzymes in bleaching of textiles is to remove the cellulose layer which was fixed at the textile fabric to protect it against breakage during spinning and weaving. This removal of the cellulose layer is accomplished by an oxidation catalyzed by the used enzyme. In this reaction, oxygen is consumed from which hydrogen peroxide is formed as a reaction product. This hydrogen peroxide can be used to bleach the textile. To control the process, the following precautions must be taken:

Dependent on the pH value for which the enzymes obtain their optimal activity, the detection of hydrogen peroxide can be done with the developed sensor system for pH values higher than 10.5 and by means of the additional FIA setup for pH values lower than 10.5.

The bleaching bath needs a system for purging air into the bath continuously, because oxygen is consumed in the process, otherwise the dissolved oxygen concentration falls down. This also causes a decrease of the hydrogen peroxide concentration which has a negative influence on the quality of the bleached product.

Bleaching with enzymes has the benefit that no hydrogen peroxide has to be added (it is made in situ) and that two processes (removal of the cellulose layer and the bleaching itself) can be done at one machine and at the same time. However, the drawbacks mentioned in the above paragraph prevail and are still a big obstacle for using enzymatic bleaching on a large scale.

REFERENCES

1. L. X. Kong, R. A. Flatfoot, and X. Wang. Text. Res. J. 66:30 (1996).
2. N. K. Lauge. Text. Chem. Color. 29:23 (1997).
3. M. Arifoglu, W. N. Marmer, and C. M. Can. Text. Res. J. 6:319 (1990).
4. S. Veleva and A. Georgieva. Reaction Kinet. Catal. Lett. 58:177 (1996).
5. B. A. Evans, L. Boleslowski, and J. E. Boliek. Text. Chem. Color. 29:28 (1997).

6. G. Rösch. Text. Praxis Int. *4*:384 (1988).
7. M. Weck. Text. Praxis Int. *2*:144 (1991).
8. P. Ney. Text. Praxis Int. *11*:856 (1974).
9. L. Jimenez, E Navarro, and J. L. Ferrer. Afinidad *54*:38 (1997).
10. S. P. Parker (ed.), *Encyclopedia of Chemistry*, McGraw-Hill Book Co., New York, 1982.
11. M. Pourbaix (ed.), *Atlas d'équilibres électrochimiques*, Gauthier Villars, Paris, 1963.
12. W. Ney. Text. Praxis Int. *10*:1392 (1974).
13. D. R. Lide (ed.), *Handbook of Chemistry and Physics, edition 71*, Chemical Rubber Publishing Company, London, 1990.
14. F. Franks (ed.), *Water: A Comprehensive Treatise, Volume 1*, Plenum Press, New York, 1972.
15. P. Wurster. Textilveredlung *22*:230 (1987).
16. J. Dannacher and W. Schlenker. Textilveredlung *25*:205 (1990).
17. A. J. Bard (ed.), *Encyclopedia of Electrochemistry of the Elements, Volume II*, Marcel Dekker, New York, 1974.
18. J. Dannacher and W. Schlenker. Text. Color. Chem. *28*:24 (1996).
19. K. Schliefer and G. Heideman. Melliand Textilberichte *11*:856 (1989).
20. J. Polster and H. Lachmann (eds.), *Spectrometric Titrations*, VCH, Weinheim, 1989.
21. P. Ney. Text. Praxis Int. *11*:1552 (1974).
22. HENKEL GmbH, DIN Sicherheitsdatenblatt 003167DE (1988), 1-2.
23. HENKEL GmbH, DIN Sicherheitsdatenblatt 001251DE (1987), 1.
24. HENKEL GmbH, DIN Sicherheitsdatenblatt 005901DE (1981), 1-2.
25. M. Jola. Melliand Textilberichte *1*:931 (1980).
26. W. C. Schumb, C. N. Satterfield, and R. L. Wentworth (eds.), *Hydrogen Peroxide*, Reinhold, New York, 1955.
27. A. C. Cumming and S. A. Kay (ed.), *Quantitative Chemical Analysis*, Oliver and Boyd, Edinburgh, 1945.
28. F. Oehme and K. Laube. Melliand Textilberichte *6*:616 (1965).
29. K. Laube and H. Zollinger. Melliand Textilberichte *78*:727 (1965).
30. J. Eisele and S. Hafenrichter. Melliand Textilberichte *7*:756 (1954).
31. E. Pungor (ed.), Ion-selective Electrodes Conference, Budapest, 1977.
32. H. W. Nurnberg (ed.), *Electroanalytical Chemistry*, Wiley, London, 1974.
33. P. Westbroek, B. Van Haute, and E. Temmerman. Fresenius J. Anal. Chem. *354*: 405 (1996).
34. P. Westbroek, E. Temmerman, and P. Kiekens. Melliand Textilberichte *79*:62 (1998).
35. Y. Cai and S. K. David. Text. Res. J. *67*:459 (1997).
36. P. Westbroek, E. Temmerman, and P. Kiekens. Anal. Commun. *35*:21 (1998).

6

Surface Features of Mineral-Filled Polypropylene Filaments

BRIAN GEORGE   School of Textiles and Materials Technology, Philadelphia University, Philadelphia, Pennsylvania

SAMUEL HUDSON and MARIAN G. McCORD   Department of Textile Engineering, Chemistry, and Science, College of Textiles, North Carolina State University, Raleigh, North Carolina

I. INTRODUCTION

Polypropylene is considered to be a material that is easy to work with because of its relatively low melting temperature and because it can be easily converted into filament form through melt spinning. Therefore, it is an ideal material to study the effects of mineral fillers on the surface properties of filaments. Polypropylene has been utilized for a number of years as a matrix material for composites and, as such, has been combined with mineral fillers. However, there have been few attempts to create mineral-filled polypropylene or other thermoplastic filaments. Currently, only Kim and White [1] have published literature concerning mineral-filled filaments, but their report focuses mostly on the orientation of the talc filler in polystyrene filaments. Research in this area has also been completed by the groups of George, Hudson, McCord, and Qiu, as well as Jack and Qiu, all at North Carolina State University.

II. MATERIALS

As mentioned previously, the filaments consist of polypropylene, specifically Pro-Fax MI 40 heat-stabilized polypropylene, which was supplied in pelletized form from Himont Incorporated. The fillers consist of talc, wollastonite, calcium carbonate, and titanium dioxide, with their properties listed in Table 1. The talc, Jetfil 700C, was supplied by Luzenac America, Inc. The talc particles, shown in Fig. 1, have a discotic morphology and a fairly smooth and featureless surface, although there are some particles with irregular surfaces. The average size of the talc particles as determined by Luzenac is 2 μm across the face and 0.2 μm in height, although the scanning electron microscopic (SEM) image exhibits many particles that are of different dimensions.

NYCO Minerals Inc. provided NYAD 400 wollastonite, which has a needle-like morphology, as exhibited in Fig. 2. NYCO states that the average size of these particles is 8.5 μm in length and 2.8 μm in diameter. However, as depicted in the image, the size of these particles varies greatly, with some very large particles contained in this figure. Surface features of the wollastonite particles vary from smooth and featureless to rough and irregular. The rough irregular surfaces are most likely the result of the mining and production of the wollastonite.

The calcium carbonate was obtained from Fisher Chemicals. These particles have the morphology of parallelepipeds, or cubes, for the most part, although the

TABLE 1 Properties of Fillers

Filler	Morphology	Average dimensions[a] (μms)	Aspect ratio
Talc	Discotic	2 (d) × 0.2 (t)	10
Wollastonite[b]	Needlelike	8.5 (l) × 2.8 (d)	3
Calcium carbonate	Parallelepiped	9.1 × 6.1 × 3.2	2.8
Titanium dioxide[c]	Spherical	0.15 (d)	1

[a] l = length, d = diameter, t = thickness.
[b] Nyco literature.
[c] Du Pont literature.
Source: Ref. 1.

FIG. 1 Talc particles.

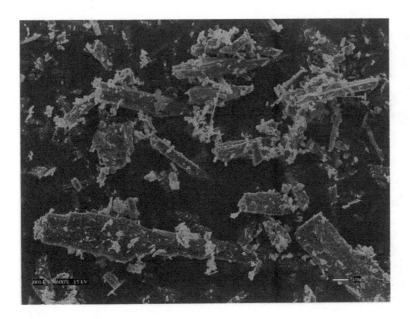

FIG. 2 Wollastonite particles.

FIG. 3 Calcium carbonate particles.

FIG. 4 Titanium dioxide particles.

morphology varied from cubical to rectangular. The size of the particles ranged from less than 5 μm to approximately 20 μm in length, as measured from the images. The surfaces of the particles are mostly smooth and featureless, whereas the edges are sharp, which can lead to stress concentrations in the composite filaments. As displayed in Fig. 3, the calcium carbonate particles have a tendency to form agglomerations.

Titanium dioxide, in the form of Ti-Pure R102, was supplied by Du Pont Chemicals. These spherically shaped particles are relatively uniform with a diameter of 0.15 μm, as determined by Du Pont. These particles have no noticeable surface irregularities. As displayed in Fig. 4, the titanium dioxide particles have a tendency to adhere to each other and form agglomerations.

III. FILAMENT PRODUCTION

Prior to producing composite filaments, composite rods were produced. These were produced with the use of an Atlas Laboratory Mixing Extruder which simultaneously melted the polypropylene while mixing it with the powdered filler. The rods were then chipped up into pellets. These composite pellets were then melted and extruded into filament form with a Bradford University Research Ltd. piston extruder, using a spinnerette size of 0.6 mm. The filaments were only drawn slightly upon extrusion and should be considered "as spun" filaments. Filaments with volume fractions (V_f) of filler of 0.00, 0.05, 0.10, 0.15, and 0.20 were produced for viewing.

IV. SURFACE FEATURES OF THE FILAMENTS

After production, these filaments were mounted and sputter coated with a gold–palladium alloy and stored in a vacuum dessicator. They were viewed with a Philips 505 scanning electron microscope utilizing an accelerating voltage of 15 kV. Image acquisition was performed with a Northgate Computer Systems Ultra utilizing Digital Scan Generator 1 Plus, version 1.43. The filler particles were also viewed in this manner, with the exception of the calcium carbonate, which was viewed with a Hitachi S-3200N environmental scanning electron microscope, with an accelerating voltage of 20 kV, interfaced with a Polaroid 545i camera using Type 55 film.

A. Polypropylene

As shown in Fig. 5, the pure polypropylene filament does not have any distinguishing characteristics, but, instead, has a mostly smooth surface. However, some minor pitting of the surface is present. These pits are attributed to air trapped in the melt during extrusion. Due to the piston type of extruder, air that is con-

FIG. 5   Polypropylene filament.

tained in the barrel during the extrusion process cannot escape except through the spinnerette. During extrusion, the air forms bubbles in the melt, with those bubbles near the surface of the filament collapsing upon filament formation, resulting in small pits. Additionally, as would be expected, the diameter of this filament is constant.

B.   Polypropylene–Talc

At a volume fraction of talc of 0.05, the surface of the filament is still smooth, as depicted in Fig. 6. However, unlike the pure polypropylene filament, this filament has some variations in fiber diameter, in the form of protuberances. Also, the filament depicted has few holes or pits, much less than the unfilled polypropylene filament. The talc displayed a tendency to agglomerate, and these protrusions are believed to be caused by agglomerations located near the surface of the filament.

When the talc V_f was increased to 0.10, the protuberances due the talc aggregations are more pronounced and evident (Fig. 7). The conglomerations of talc also appear to be less spherical and more of a random morphology. Also, these protrusions do not always gradually alter the diameter of the filament, as with the lower volume fraction of talc; instead, they sometimes abruptly disrupt the continuity of the filament. The surface of this filament has slightly more pitting

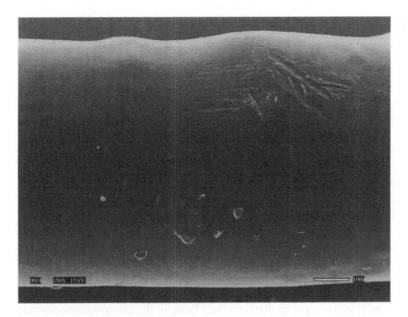

FIG. 6 Polypropylene–talc, $V_f = 0.05$.

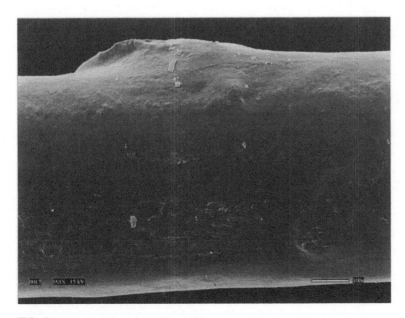

FIG. 7 Polypropylene–talc, $V_f = 0.10$.

than the $V_f = 0.05$ filament, but it still has less than the polypropylene filament. This image depicts a filament that is not as smooth as either the $V_f = 0.05$ or pure polypropylene filaments. During cross-sectional microscopy, it was determined that the talc was generally well distributed throughout the cross section of the filament. Therefore, it is reasonable to conclude that this rougher surface is due to more talc particles located near or on the surface of the filament.

This theory is borne out when the surface of the $V_f = 0.15$ filament is examined (Fig. 8). This filament is similar to the $V_f = 0.10$ filament in that there are rather large agglomerations which disrupt the smoothness of the surface. As the volume fraction of talc has increased, so has the roughness of the surface; however, this particular filament has ridges in its surface, as opposed to the filaments containing less talc, which have relatively smooth surfaces in comparison. The cause of these ridges is unknown, but it appears that they traverse the length of the filament and are visible on both the upper and lower portions of the filament. However, it appears that the very top of the filament is smoother, which would indicate that these ridges are only located on part of the circumference of the filament, as opposed to having ridges around the entire circumference of the filament. This specimen does not appear to have many air bubbles, but those that are evident appear to be slightly larger in diameter than those in the lower-volume-fraction

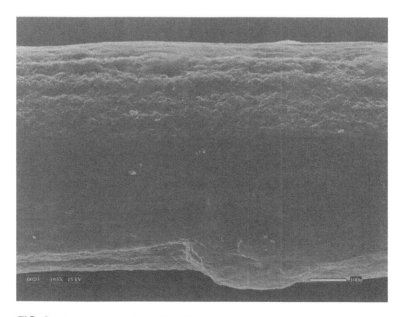

FIG. 8 Polypropylene–talc, $V_f = 0.15$.

filaments. It is possible that some bubbles are obstructed by the ridges on the surface.

The $V_f = 0.20$ filament has a similar surface to the $V_f = 0.15$ talc-filled filament, but in this image (Fig. 9), it can be seen that these lengthwise ridges do not travel around the entire circumference of the filament, as there are some areas without ridges in the lower portion of the filament. The same spinnerette was utilized for the spinning of all the filaments, but as noted earlier, the talc had a tendency to agglomerate as well to adhere to surfaces with which it came into contact. Some residual talc was found in the mixer–extruder, as well as in the Bradford extruder, when it was disassembled, which leads to the hypothesis that perhaps these ridges are due to some talc adhering to the diameter of the spinnerette and thus disrupting the flow of the melt during extrusion. No large protrusions from the filament surface are evident in this image, although several of the ridges appear to contain large protuberances. As with the other talc-filled filaments, these protuberances are most likely talc agglomerations that are located near the surface of the filament. Some bubble remnants are visible in both the smooth and rough areas of this filament. These remnants appear to be larger and slightly more numerous than those found in the lower-volume-fraction filaments.

The talc-filled filaments present an interesting scenario as the volume fraction of talc increases. The $V_f = 0.05$ filament has a relatively smooth and featureless

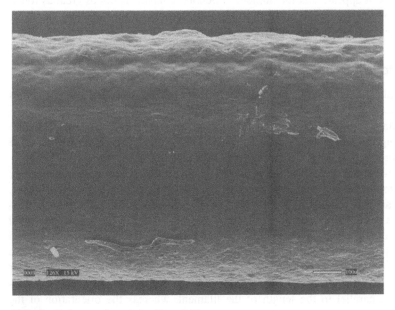

FIG. 9 Polypropylene–talc, $V_f = 0.20$.

surface, although there are some diameter variations most likely caused by agglomerations of talc located in close proximity to the surface of the filament.
As the volume fraction increases, the agglomerations become more pronounced,
perhaps because larger agglomerations or more numerous agglomerations are
forming due to the increased amount of talc contained in these filaments. However, the $V_f = 0.10$ filament is still relatively smooth, especially when compared
to the higher-volume-fraction filaments. The $V_f = 0.15$ and $V_f = 0.20$ filaments
have some surface areas which are extremely textured in the form of ridges.
Because these ridges are not located throughout the entire circumference of the
filaments, it is believed that they are caused by partial obstruction of the spinnerette, which disrupted the flow of the melt during extrusion. This is a likely scenario because the ridges are fairly continuous and do not appear to deviate from
their longitudinal direction along the filament length. If the spinnerette had talc
adhering to it in certain areas in large enough quantities to obstruct the flow, the
filament would have depressed channels along its length, similar to the images
of the 0.15 and 0.20 volume fraction of talc filaments. These ridges or channels
may be avoidable if the spinnerette is examined for obstructions or cleaned prior
to every extrusion. Overall, the ability of the talc to adhere to surfaces that it
came into contact with, including other talc particles, may deem it unsuitable for
use if smooth featureless filament surfaces are desired. Additionally, the pitting of
the specimens seems to increase in both numbers and size as the volume fraction
increases. It appears that the number of bubble remnants is not as great as found
in the polypropylene filament, but the different topography of these filaments
may obscure some remnants from view.

C. Polypropylene–Wollastonite

The wollastonite-filled filament with a volume fraction of 0.05 wollastonite has
a surface that is similar to the talc-filled filament of the same volume fraction.
Both filaments are relatively smooth, although there are some variations in diameter along the length of the filament in the image (Fig. 10). However, there are
some protrusions on the surface of the filament, similar to those on the higher-
volume-fraction talc-filled filaments. Most likely, these are agglomerations of
wollastonite located near the surface of the filament, due to their size. However,
it appears that some wollastonite particles are located on the surface of the filament as well. These mostly appear as white specks, but in the lower right-hand
corner of Fig. 10, it appears that there is a long wollastonite particle located
partially on the surface of the filament. This particle is much longer than the
average stated by NYCO, but in Fig. 2, some of the particles have much greater
lengths than the average. This particle can be seen to be oriented so that its
length is parallel to the length of the filament, whereas the orientation of the
other filaments cannot be determined, due to their small size in relation to the

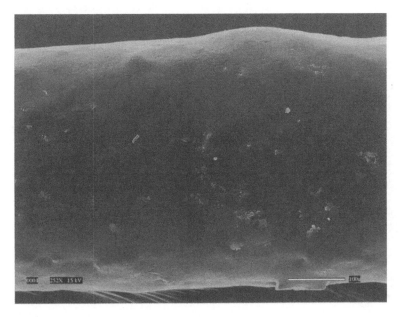

FIG. 10   Polypropylene–wollastonite, $V_f = 0.05$.

size of the filament. When compared to the polypropylene and $V_f = 0.05$ talc filaments in terms of bubble remnants, this filament is more similar to the polypropylene specimen, as it contains many small bubble remnants.

The $V_f = 0.10$ wollastonite filament has a much different appearance than the $V_f = 0.05$ filament, as shown in Fig. 11. Whereas the lower content filament is mostly smooth with some small protrusions, Fig. 11 depicts a filament which has a rough irregular surface. This filament has significantly more wollastonite agglomerations located near the surface, which are responsible for the change in surface topography between the different volume fractions of this mineral. The size and morphology of these agglomerations are not uniform, which also increases the texture of this filament. This filament appears to have had more air trapped near the surface of the filament than the $V_f = 0.05$ specimen evidenced by the more noticeable depressions in the surface. Additionally, it appears that some of these remnants are linked to each other by shallow channels in the surface. These channels are not always straight; some appear to be curved and form geometric patterns on the surface. The cause of these channels is unknown, but it may be due to turbulence in the melt due to the presence of the wollastonite particles.

The $V_f = 0.15$ filament continues the trend of increasing surface texture with increasing volume fraction of wollastonite (Fig. 12). This filament contains nu-

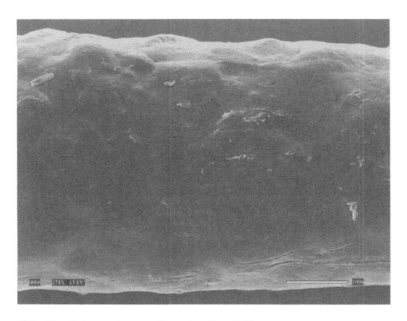

FIG. 11 Polypropylene–wollastonite, $V_f = 0.10$.

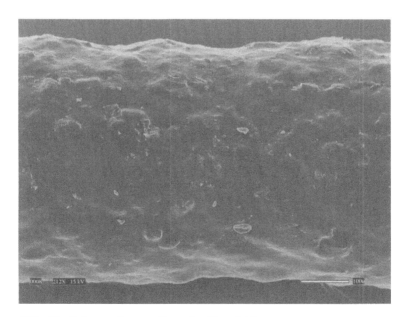

FIG. 12 Polypropylene–wollastonite, $V_f = 0.15$.

merous protrusions of various sizes and morphologies, with no detectable pattern. As a result, the topography of this filament appears to be quite rough. Additionally, as with the $V_f = 0.10$ specimen, it appears that some wollastonite particles may be located on the surface of the filament, which would increase the texture of this filament. However, this specimen does have as many bubble remnants as the $V_f = 0.10$ filament. Upon close inspection, there are several remnants, with many of them located between protrusions, so it is possible that the surface topography is obscuring more remnants.

The $V_f = 0.20$ specimen appears to be an amalgamation of the earlier filaments (Fig. 13). As with the other filaments containing wollastonite, it appears that some filler particles are located on the surface of this filament. In fact, it appears that there are some large filler particles oriented with their lengths fairly parallel to the length the filament, similar to the $V_f = 0.05$ filament. It is difficult to discern the orientation of the smaller wollastonite particles, due to their diminished size, but because of their smaller size in relation to the filament size as well as the lack of drawing, these filler particles may not have any orientation. This specimen, as with the other high-volume-fraction specimens also suffers from many agglomerations near the surface of the filament. They are differentiated from the large filler particles by their morphology, with the aggregates generally having more amorphous shapes. Pitting of this filament is evident, and similar to the $V_f =$

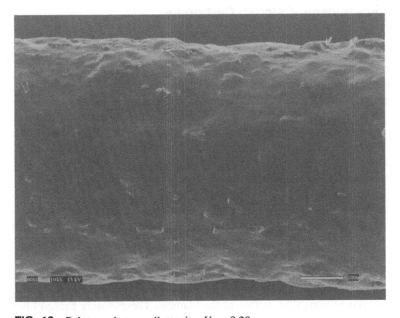

FIG. 13 Polypropylene–wollastonite, $V_f = 0.20$.

0.10 specimen, some of these bubble remnants seem to be linked by shallow depressions in the surface. Once again, these depressions are thought to be due to flow turbulence of the melt as a result of the presence of the wollastonite particles. The area viewed on this specimen does not appear to be as textured as the $V_f = 0.15$ filament, and it is possible that these depressions only form in untextured areas.

As the volume fraction of wollastonite increases, the protrusions on the suface of the filaments become more pronounced, up to a volume fraction of 0.15 wollastonite. Due to the increase in number of protrusions, the filaments have rougher surfaces as the volume fraction increases. However, the $V_f = 0.20$ specimen has a less textured surface than the $V_f = 0.15$ specimen. The number of air-bubble remnants also appears to increase with the volume fraction, although the $V_f = 0.15$ filament seems to have less remnants than both the $V_f = 0.10$ and $V_f = 0.20$ specimens. This may be an illusion due to the increased number of protrusions shrouding the remnants. An interesting feature of both the $V_f = 0.10$ and $V_f = 0.20$ filaments is that they both have shallow channels in smooth areas of the surface that link some of the pits and form amorphous shapes. These are believed to be due to the wollastonite particles disrupting the flow of the polypropylene matrix during extrusion. Overall, these filaments do not have smooth surfaces, particularly at levels of wollastonite above 0.05.

D. Polypropylene–Calcium Carbonate

The $V_f = 0.05$ polypropylene–calcium carbonate filament is much different in surface features than the comparable talc- and wollastonite-filled filaments. The first feature noticed when examining this specimen is that it contains several agglomerations which protrude through its surface (Fig. 14). These protrusions are discrete in that they are not linked together, nor are they located close enough together to give an overall impression of a rough surface. Instead, the filament seems to have a smooth surface interrupted by protuberances. The second main feature of this filament is that some of the filler particles appear on the surface, in the form of small white specks. Another feature of this particular specimen is that it contains many small pits on the surface, remnants of collapsed air bubbles, as mentioned previously.

The $V_f = 0.10$ specimen in Fig. 15 has a similar appearance to the lower-volume-fraction filament, except for the increased number of protuberances and filler on the surface. This specimen has many more protrusions than the lower-volume-fraction filament, giving it a much rougher appearance. As with the $V_f = 0.05$ volume fraction filament, there are many filler particles on the surface of the filament. In some instances, it appears that some of the agglomerations are actually breaking through the surface of the filament. These closely packed filler particles in some surface agglomerations resemble large filler particles entrapped

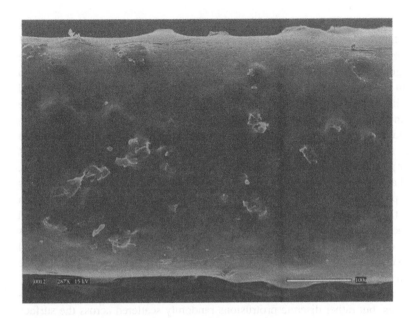

FIG. 14 Polypropylene–calcium carbonate, $V_f = 0.05$.

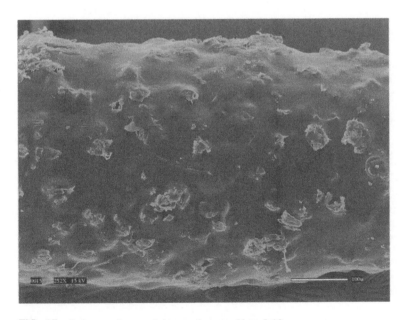

FIG. 15 Polypropylene–calcium carbonate, $V_f = 0.10$.

in the surface of the filament. However, in other areas of the specimen, it does appear that extremely large filler particles are projecting through the surface of the filament. In fact, there are some holes on the surface of the filament that are too large and irregularly shaped to be due to air. These presumably are due to calcium carbonate particles that were not securely held by the polypropylene and were removed during handling. As with many of the other filaments, this one has many bubble remnants on its surface. As depicted in Figs. 16 and 17, the $V_f =$ 0.15 and $V_f = 0.20$ polypropylene–calcium carbonate filaments are very similar in surface topography to the $V_f = 0.10$ filament. They all have many protrusions of filler particles and agglomerations breaking through the surface of the filament, as well as many conglomerations situated close to the surface which form bulges in the specimens. They also show pitting of the surface due to escaping air from the melt. The main difference between these filaments appears to be the increased cragginess of the surface due to the increased number of filler agglomerations located directly under the surface of the filaments. The amount of filler located on the surface of the filaments seems to be equal among the three higher volume fractions of calcium carbonate.

Overall, the calcium-carbonate-filled specimens appear to contain no continuous ridges, but rather discrete protrusions randomly scattered across the surface of the filaments. As the volume fraction of filler increases, the number of protru-

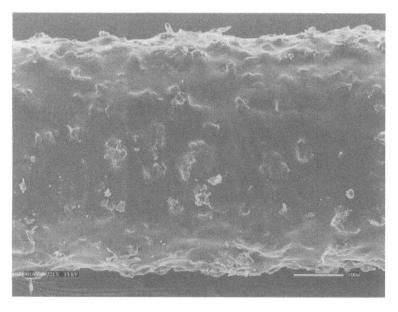

FIG. 16   Polypropylene–calcium carbonate, $V_f = 0.15$.

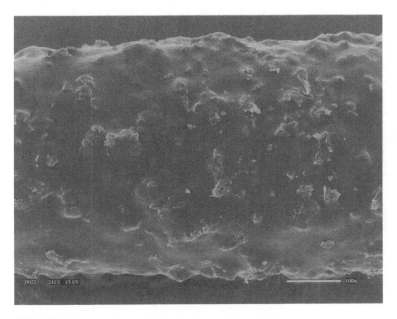

FIG. 17   Polypropylene–calcium carbonate, $V_f = 0.20$.

sions also increases. Many filler agglomerations as well as individual filler particles are contained on the surface of these filaments, with the amount of both
increasing between the $V_f = 0.05$ and $V_f = 0.10$ volume fractions. However, at
the upper three filler concentrations, there does not seem to be a major increase
in the amount of filler particles and agglomerations on the surface. The only
difference among these specimens seems to be the amount of protuberances on
the surface, giving each successively greater volume fraction an increased surface
topography, resulting in a rougher surface. All of the fillers also exhibit remants
of air bubbles on their surfaces, which is due to air trapped in the melt during
extrusion.

E. Polypropylene–Titanium Dioxide

Titanium dioxide, a whitener, as a filler provided some of the most challenging
problems during the mixing and extruding processes. The titanium dioxide had
a tendency to adhere to surfaces with which it came into contact. As a result,
both the mixer and extruder needed to be disassembled and cleaned after use
with titanium dioxide. The lowest-volume-fraction specimen, $V_f = 0.05$, much
greater than usually used in the production of filaments, is exhibited in Fig. 18.
At this level of titanium dioxide, the filament exhibits a fairly smooth surface.

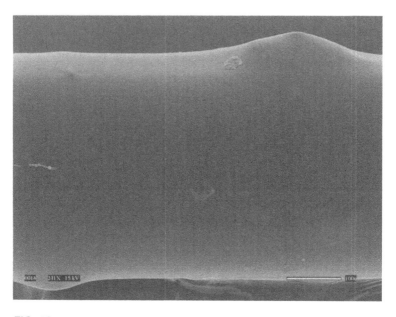

FIG. 18   Polypropylene–titanium dioxide, $V_f = 0.05$.

However, some agglomerations under the surface are evident by the protrusions. These protrusions seem to start diverging from the plane of the surface gradually, instead of immediately like some other fillers, such as calcium carbonate. Some pitting, as witnessed in other filaments, is also evident, but these are fairly small. Overall, this filament seems to have a smooth appearance.

When the $V_f = 0.10$ titanium dioxide filament is viewed (Fig. 19), it is evident that the increased level of filler has led to an increased amount of protrusions. However, as described earlier, most of these protrusions gradually rise above the surface of the filament. This specimen has variations in its diameter, which was not observed in the $V_f = 0.05$ filament. This filament suffers from more pitting from air-bubble remnants than the lower-volume-fraction specimen, and it has shallow depressions in the surface which intersect with pits and other depressions to give the surface a scaly appearance. As mentioned previously, these depressions may be due to the filler particles disrupting the flow of the matrix during filament extrusion. The white specks on the surface of this filament may be due to dust or titanium dioxide conglomerations breaking the surface of polypropylene.

Figure 20 depicts the filament containing $V_f = 0.15$ titanium dioxide. This filament differs from the lower volume fractions in that approximately equal amounts of the protrusions rise abruptly from the filament surface as rise gradually. However, it appears that the type of protrusion depends on its size, with the

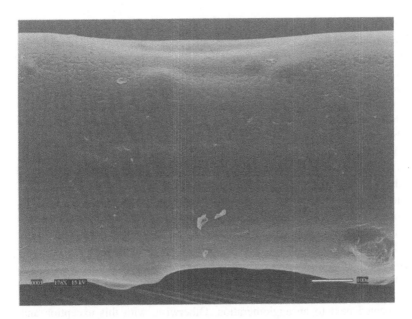

FIG. 19 Polypropylene–titanium dioxide, $V_f = 0.10$.

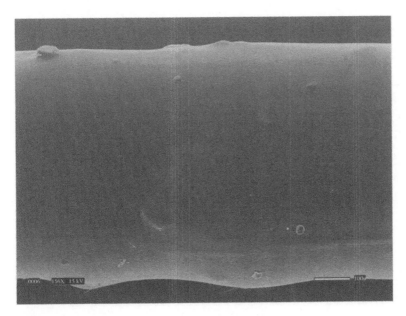

FIG. 20 Polypropylene–titanium dioxide, $V_f = 0.15$.

larger protuberances rising gradually from the filament surface. This filament also suffers from pitting due to collapsed air bubbles, but it does not appear to be as evident as on the $V_f = 0.10$ filament. Also, although there are some shallow grooves in the surface, these are not as prevalent as on that the filament, nor do these grooves form patterns of any type. It appears that some small agglomerations of titanium dioxide can be found breaking the surface of the polypropylene, although these could be dust particles as well. With the exception of the protrusions, this filament has a very regular and smooth surface.

The $V_f = 0.20$ filament is depicted in Fig. 21. This filament has some similar features to the previously examined polypropylene–titanium dioxide filaments, while also having some unique surface properties. First, as with the other titanium-dioxide-filled specimens, the surface of this filament is relatively smooth. The filament does not have many agglomerations near the filament surface, nor does it have any sharply rising protrusions. Some protuberances from conglomerations of titanium dioxide located near the surface are evident, but these protrude gradually from the surface and form a hemisphere, reminiscent of the smooth spherical titanium dioxide particle. This filament does not seem to have any major diameter variations, but there is a major depression in the filament in the lower portion located next to an agglomeration. Otherwise, with this exception and

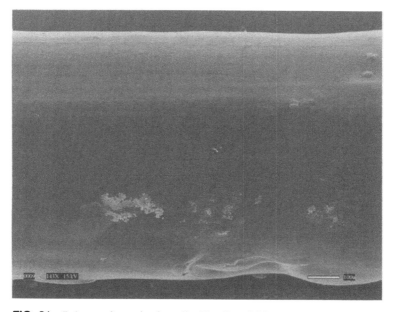

FIG. 21 Polypropylene–titanium dioxide, $V_f = 0.20$.

some protuberances, the surface of this filament is relatively unaffected by the titanium dioxide.

The filaments containing titanium dioxide appear to be the least affected by the addition of the filler in terms of surface morphology. Although protrusions are evident, many of these do not rise sharply from the filament surface, but they gradually increase the diameter of the filament in that one area. Many of these protuberances are hemispherically shaped, reminiscent of the smooth spherical titanium dioxide particle, although there are some with sharp edges that rise abruptly from the surface. Overall, these filaments appear to be the smoothest at all volume fractions of filler when compared with the other filaments.

V. CONCLUSION

Mineral fillers can be utilized to alter the properties of polypropylene filaments. As depicted in the many images, the surfaces of filaments containing filler are usually altered from that of the pure polypropylene filament. The different fillers result in differing filament surface topographies. The talc-filled filaments at the lower volume fractions have smooth surfaces interrupted by agglomerations. As the volume fraction of talc increases, the surface becomes ridged in some areas. However, these ridges may be a by-product of talc adhering to the surface of the spinnerette during extrusion. Above $V_f = 0.05$ talc, the filaments suffer from large agglomerations of talc, which would give the filaments a bumpy texture along their lengths. As the volume fraction of wollastonite increases, the surface topography becomes rougher. This is due to the many protuberances caused by agglomerations of wollastonite located near the surface. Additionally, some wollastonite particles can be seen located on or directly under the surfaces of the filaments. Calcium carbonate filaments are similar to those filled with wollastonite in terms of surface features. As the volume fraction of calcium carbonate increases, the surface texture also increases, to give these filaments a rough feel. This is due to the many agglomerations of filler located on or under the surface of the filaments. Some filler particles can be seen on the filament surfaces, whereas agglomerations of filler particles are located directly under the surface, forming protrusions on the surface of the specimens. The titanium dioxide filaments are perhaps the least affected by the presence of the filler. These specimens are relatively smooth and featureless with the exception of some protuberances. However, many of these protuberances are relatively smooth themselves, giving the filaments a smooth texture. In terms of the surfaces, the calcium-carbonate-filled and wollastonite-filled filaments will probably have the dullest appearances due to the rough topography of the filaments, which will diffuse incident light. The talc-filled filaments at the lower volume fractions, $V_f = 0.05$ and V_f 0.10, will most likely have a mixture of diffuse and specular reflections due to their relatively smooth surfaces. However, at the upper volume fractions, $V_f = 0.15$ and

$V_f = 0.20$, these filaments will likely have more diffuse light reflection than specular reflection due to the increase in surface texture. As a result, these filaments will appear duller than the lower-volume-fraction talc-filled filaments. The titanium-dioxide-filled filaments with their relatively smooth surfaces will have the greatest amounts of specular reflection of all the filaments, with the exception of the pure polypropylene filament.

The different mineral fillers utilized in producing these mineral-filled filaments give the filaments different surface properties. As a result, some filaments will have smoother surfaces and, thus, a nicer hand than others, whereas some will have a duller appearance than others. Overall, these filaments differ from the unfilled polypropylene in that they all have some sort of different surface topography. The most common feature of all the mineral-filled filaments is the presence of protuberances. These are a result of filler or agglomerations of filler located near the surface. In some filaments, actual filler particles are located on the surface of the filament as well. Other features include filament variations, ridges, and surface patterns. As a result of the inclusion of mineral fillers, interesting surface features can be obtained, which may be useful for different applications that rely on fiber surface features.

REFERENCE

1.   K. J. Kim and J. L. White. J. Non-Newtonian Fluid Mech. 66:257 (1996).

7

Inorganic Fibers

ANTHONY R. BUNSELL and MARIE-HÉLÈNE BERGER   Centre des
Matériaux, Ecole des Mines de Paris, Evry, France

I. INTRODUCTION

Ceramics as fibers seems at first view to be a strange concept. We are used to thinking of ceramics as being brittle, inflexible bulk materials with highly crystalline structures, not as flexible filaments which can be woven and turned into complex shapes. Nevertheless, natural ceramic fibers, such as asbestos fibers and rock wool, have been used for many decades. Man-made glass fibers have been available since the mid-1930s and interest in advanced fibers for reinforcement was stimulated by the development of carbon fibers made from organic precursors in the mid-1960s. The surfaces of glass fibers have to be protected by a size so as to prevent their deterioration by abrasion and by the environment and this size also acts as a lubricant and coupling agent to the matrix. The surfaces of carbon fibers are treated so as to improve interfacial bonding with the resin matrix and a coupling agent can also be added to the fiber surface. Synthetic ceramic fibers were developed in the 1960s, although they had large diameters and could not be woven. Today, the main interest in this type of fiber is for the reinforcement of titanium, and for this, the surface of the fiber has to be protected by a complex coating to avoid degradation during composite manufacture. Some continuous ceramic fibers with small diameters of around 20 μm were developed in the 1970s, and since 1980, a number of small-diameter ceramic fibers have been developed, some of which can be woven and which offer the possibility of reinforcing bulk ceramic matrices for use at very high temperatures. The inclusion of ceramic fibers in a brittle ceramic or vitroceramic matrix implies a control of the fibers' surfaces, as the bonding of the fibers with the matrix has to be carefully controlled so as to maximize the control of crack propagation.

The flexibility of fibers is determined by their fineness so that materials, which in bulk form appear very inflexible, can be made into fine filaments which can then behave like textile fibers. An everyday example is glass fibers. The properties of the glass used for making glass fibers are very similar to those of bulk glass used, for example, in windows. However, glass fibers can be woven into cloth in much the same way as are nylon fibers and similar looms are used for the two. The flexibility of glass fibers comes from their small diameter, which is usually between 5 and 20 μm and this, even with a Young's modulus which is similar to that of aluminum (i.e., around 70 GPa) confers great suppleness on them. The flexibility of a cylinder increases as its diameter decreases as a function of its radius to the fourth power. The Young's moduli of most synthetic textile fibers are usually low, not more than about 5GPa; however, fine filaments of carbon are produced with Young's moduli up to 800 GPa which approaches that of diamond (1200 GPa), which is the stiffest material known. The Young's modulus of steel is 210 GPa. Therefore, the hardness or stiffness of ceramics is not an insurmountable problem in converting them into fibers. Fineness lends flexibility to even the stiffest of materials.

Ceramics, however, are brittle and show much scatter in their mechanical properties. This is true of many fibers, and even if a stiff material can be made flexible in filament form, its intrinsic properties will control its ultimate behavior. A glass fiber can be bent easily but only down to a minimum radius of curvature, as the tensile stresses developed in the convex surface will eventually reach the tensile failure stress of the glass and the fiber will break. Glass is an amorphous material which does not deform plastically at room temperature, so that even in the form of a flexible fiber, it will break by the same elastic processes as bulk glass. This depends on the presence of defects on the fiber surface or within its volume. A ceramic material is crystalline and shows great scatter in strength because of defects which are distributed throughout its volume and this can be related to processing faults and the large sizes of the grains which make up its microstructure. The long, fine form of a filament, however, demands the control of a much finer microstructure than that usually found with bulk ceramics, and as the diameter of the fiber is reduced, the ratio of surface area to volume increases, thus reducing the importance of flaws in the volume. Ceramic fibers are therefore stronger than their counterparts in bulk form; however, like glass and carbon fibers, they remain elastic at room temperature and fail in a brittle manner.

The fibers discussed here are used for the reinforcement of composite materials. The glass fibers are used to reinforce organic resin systems, as are usually the carbon fibers. Ceramic fibers are used in resin systems to modify their dielectric properties, but our main concern here is their use as reinforcements in metal or ceramic matrices, for use at high temperatures.

The first full-scale commercial production of glass fibers began in the United States in the 1930s by two companies which eventually fused to become Owens Corning, which is still the largest glass fiber manufacturer. The original uses for the glass fibers were in the form of a felt as industrial filters, but the production of thermosetting resins in the 1940s led to glass fibers becoming the most widely employed reinforcement for composite materials and these have seen a continuous increase in their use since the 1950s. The fibers have to be protected by a surface size which also acts as an adhesive to control the fiber–matrix interface as well as a lubricant to facilitate fiber processing, and when the fibers are used in a chopped form, the size also contains an antistatic agent to permit uniform distribution of the fibers during composite manufacture.

Carbon fibers were first developed by Edison at the beginning of the twentieth century by the pyrolysis of bamboo fibers and used for several years as the filaments in the first electric light bulbs, only later to be replaced by less brittle tungsten filaments. Research in the 1950s, primarily in the United States, into the production of strong carbon fibers as reinforcements concentrated on the pyrolysis of regenerated cellulose fibers and led to the development of carbon fibers which, today, are mainly used in carbon–carbon composites for use as reentry heat shields and as brakes. High-performance carbon fibers were eventually developed, using a purely synthetic organic precursor, in Great Britain in the mid-1960s and also in Japan at about the same time.

The earliest forms of ceramic fibers were made by chemical vapor deposition of boron or silicon carbide onto a tungsten or carbon filamentary core. The resulting fibers, which have large diameters ranging from 100 to 140 μm, are not flexible and cannot be woven or used to make complex shapes. They are used in the boron–aluminum tubular frame of the American space shuttle and, more recently, such silicon carbide fibers embedded in a titanium matrix have been considered as a future structural material for jet engines.

Ceramic fibers with diameters of the order of 10 or 20 μm based on alumina began to appear in the 1970s and later fibers, based on silicon carbide, were commercially developed in the early 1980s, which allowed ceramics to be reinforced. Such composite materials were seen as being potentially important structural materials for use at very high temperatures.

The role of the fine ceramic fibers used to toughen ceramic matrices is very different from fibers used in a resin or metal matrix. This difference arises from the nature of the matrix materials used. Resin and most metal composites exploit the lower rigidity and the ability of the matrix material to deform plastically. This, together with the greater rigidity of the reinforcements, allows stresses to be transferred to the fibers through the matrix and their extraordinary mechanical properties to be conferred to the composite structure as a whole. In contrast, ceramic matrix composites usually are composed of a very stiff matrix in which less stiff fibers are embedded. In this way, the roles, to some extent, are reversed compared to a traditional composite. The stiffness of the whole composite is primarily governed by the matrix. The role of the fibers is to mitigate against the brittleness of the matrix. The fibers must stop cracks propagating in the ceramic matrix. Two mechanisms exist for this: the debonding of the fiber–matrix interface and the bridging of cracks by fibers and their subsequent pull out. In both of these processes, the interfacial bond is of primary importance.

One of the most challenging questions exciting both producers and researchers is how to develop fibers which are stable up to very high temperatures in an oxidizing atmosphere. The environment in which the fibers are to be exposed to high temperatures is all important because, in the absence of oxygen, the choice of fiber for most applications would be that of carbon fibers. Carbon fibers have small diameters and are made with a wide range of Young's moduli; however, above 400°C, they begin to suffer greatly from oxidation. Carbon-fiber-reinforced carbon is used for reentry heat shields for space vehicles and also brakes for trucks and planes, but these are short-duration applications, and even if the carbon–carbon can resist temperatures up to 3000°C, oxidation limits its use. Glass fibers are limited to even lower temperatures and are used to reinforce resins to make composite materials which can be used at temperatures which rarely exceed 300°C.

The applications which define the temperature range required for fine ceramic fibers are those which, at present, rely on nickel-based superalloys, such as the jet engine. The upper limit of such superalloys is around 1100°C, but they can be used with ceramic insulating coatings and air cooling, through channels in the structure, in environments which are much hotter than this temperature. However, the metal must not experience temperatures higher than 1100°C and much ingenuity is put into ensuring that it does not. Design has its limits however, and if more efficient engines are to be made, materials capable of operating uncooled in air up to perhaps 2000°C are required. A more realistic goal might be 1600°C and definite advantages would be obtained with materials which were stable up to 1400°C. These materials could be required to retain their properties for periods of up to 6×10^4 h.

The ideal fiber for such applications would be both chemically and microstructurally stable up to 1600°C, have a small diameter of less than 20 μm to ensure flexibility, possess a Young's modulus around 200 GPa or higher and strength of above 2 GPa at room temperature, and retain a modulus of 150 GPa and a strength of at least 1.5 GPa at 1300°C. The fiber should also be affordable.

There are a number of routes which have been adopted to produce small-diameter ceramic fibers which could be considered for the above applications and which might fulfill the above criteria. Fine ceramic fibers based on silicon carbide are produced by the conversion of organo-silicon precursors under controlled conditions of pyrolysis. Oxide fibers, usually based on alumina, produced from sol-gels or slurries have a longer history than the silicon-carbide-based fibers and their inertness to oxidation makes them potential candidates for high-temperature use.

All of the fibers mentioned have specific advantages and drawbacks which will be discussed in this chapter.

II. GLASS FIBERS

Glass filaments have probably been formed since or before Roman times and, more recently, the production of fine filaments was demonstrated in Great Britain in the nineteenth century and used as a substitute for asbestos in Germany during the World War I [1]. In 1931, two American firms, Owen Illinois Glass Co. and Corning Glass Works, developed a method of spinning glass filaments from the melt through spinnerettes. The two firms combined in 1938 to form Owens Corning Fiberglas Corporation. Since that time, extensive use of glass fibers has occurred. Initially, the glass fibers were destined for filters and textile uses; however, the development of heat-setting resins opened up the possibility of fiber-reinforced composites, and in the years following the World War II, the fiber took a dominant role in this type of material. Today, by far the greatest volume of composite materials is reinforced with glass fibers.

Fibers of glass are produced by extruding molten glass through holes in a spinnerette with diameters of 1 or 2 mm and then drawing the filaments to produce fibers having diameters usually between 5 and 15 μm. The spinnerettes usually contain several hundred holes so that a strand of glass fibers is produced.

Several types of glass exist, but all are based on silica (SiO_2), which is combined with other elements to create speciality glasses. The compositions and properties of the most common types of glass fibers are shown in Table 1. The most widely used glass for fiber-reinforced composites is called E-glass; glass fibers with superior mechanical properties are known as S-glass.

The strength of glass fibers depends on the size of flaws, most usually at the surface, and as the fibers would be easily damaged by abrasion, either with other fibers or by coming into contact with machinery in the manufacturing process,

TABLE 1 Composition (wt%), Density, and Mechanical Properties of Various Glasses Used in Fiber Production

	Glass type				
	E	S	R	C	D
SiO_2	54	65	60	65	74
Al_2O_3	15	25	25	4	
CaO	18		9	14	0.2
MgO	4	10	6	3	0.2
B_2O_3	8			5.5	23
F	0.3				
Fe_2O_3	0.3				
TiO_2					0.1
Na_2O				8	1.2
K_2O	0.4			0.5	1.3
Density	2.54	2.49	2.49	2.49	2.16
Strength (20°C) (GPa)	3.5	4.65	4.65	2.8	2.45
Elastic modulus (20°C) (GPa)	73.5	86.5	86.5	70	52.5
Failure strain (20°C) (%)	4.5	5.3	5.3	4.0	4.5

Note: The type E is the most widely used glass in fiber production, types S and R are glasses with enhanced mechanical properties, type C resists corrosion in an acid environment, and type D is used for its dielectric properties.

they are coated with a size. The purpose of this coating is both to protect the fiber and to hold the strand together. The size may be temporary, usually a starch–oil emulsion, to aid in the handling of the fiber, which is then removed and replaced with a finish to help the fiber–matrix adhesion in the composite. Alternatively, the size may be of a type which has several additional functions which are to act as a coupling agent and lubricant and to eliminate electrostatic charges.

Continuous glass fibers may be woven, as are textile fibers, made into a nonwoven mat in which the fibers are arranged in a random fashion, used in filament winding, or chopped into short fibers. In this latter case, the fibers are chopped into lengths of up to 5 cm and lightly bonded together to form a mat, or chopped into shorter lengths of a few millimeters for inclusion in molding resins.

The structure is vitreous, with no definite compounds being formed and no crystallization taking place. A open network results from the rapid cooling which takes place during fiber production, with the glass cooling from about 1500°C to 200°C in between 0.1 and 0.3 s [2]. Despite this rapid rate of cooling, there appear to be no appreciable residual stresses within the fiber and the structure is isotropic.

Glass is set to remain the most widely used reinforcement for general composites. They are cheaper than most other relatively high-modulus fibers and, because of their flexibility, do not require very specialized machines or techniques to handle them. However, their elastic modulus is low when compared to many other inorganic fibers and the specific gravity of glass, which for E-type glass is 2.54, is relatively high. The poor specific value of the mechanical properties of glass fibers means that they are not used for structures requiring light weight as well as high strength and stiffness.

III. FIBERS ON A SUBSTRATE

A. Boron Fibers

The first fiber produced with a much increased Young's modulus compared to glass fibers was the boron fiber made by a chemical vapor deposition (CVD) technique onto a substrate which was a tungsten wire. Boron, the fifth element in the periodic table, is the lightest element with which it was found practical to make fibers. The first boron fibers were produced in the United States and then in France and in the Soviet Union at the beginning of the 1960s. They had remarkable properties, with a Young's modulus exceeding 400 GPa, as shown in Table 2, and quite extraordinary strength in compression. This latter quality was enhanced by their large diameter of 140 μm [3]. The manufacturing technique requires that each fiber be made separately and at low speed. A mixture of boron trichloride (BCl_3) and hydrogen (H_2) heated to around 1000°C passes over a

TABLE 2 Properties of Ceramic Fibers Produced by Chemical Vapor Deposition

Fiber type	Manufacturer	Trade mark	Composition (wt%)	Diameter (μm)	Density (g/cm³)	Strength (GPa)	Strain to failure (%)	Young's modulus (GPa)
SiC	Textron	SCS-6	SiC on carbon core	140	2.7–3.3	3.4–4.0	0.8–1	427
	DERA	Sigma	SiC on tungsten core	100	3.4	3.4–4.1	0.8	400–410
B	Textron		Boron on tungsten core	100–140	2.57	3–6	1	380–400

continuous tungsten filament heated by electric current. Boron is deposited preferentially in the form of nodules onto the irregularities on the tungsten filament which is produced by a wire-drawing process. The nodular structure of the fiber results in its having a surface which is not smooth and the stress concentrations produced on the surface at the intersection of two nodules can be a factor in determining the fiber strength. Other defects associated with the core–mantle interface can also limit the strength of the fiber. Figure 1 shows a cross section of a boron fiber, revealing its composite structure. The penetration of the small boron atoms into the tungsten core causes it to expand. This expansion is resisted by the boron mantle so that the core is put into compression and the mantle into tension. The cooling of the boron filament after its production induces a compressive stress into the surface.

Boron fibers proved to be good reinforcement for light metal alloys, particularly when the fiber was protected by a coating of SiC or B_4C. The American space shuttle is composed of a skeleton made up of more than 200 tubes in boron-fiber-reinforced aluminum, but the very high cost of the fiber has restricted much wider use. Most of the applications originally foreseen for boron fibers have been taken over by carbon fibers, although they are used in some sports goods. These fibers are limited in their possible temperature range, as reactions occur at the interface around 1000°C.

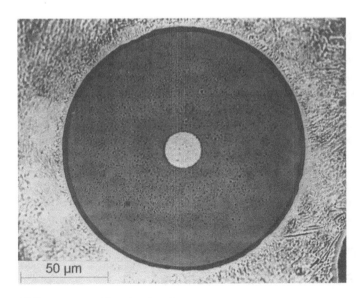

FIG. 1 Boron fiber showing the tungsten core.

B. Silicon Carbide Fibers

Silicon carbide fibers produced by a technique similar to that used to make boron fibers remain a possible reinforcement for metal–matrix composites, including titanium– and intermetallic–matrix composites. These fibers are based on materials which are readily available and the substrates can either be tungsten wire or a carbon filament, which is less expensive. Fibers of large diameters, 140 μm, are made using various chlorosilanes such as CH_3SiCl_3, which gives SiC and 3HCl. The fiber produced has a mantle of silicon carbide, which, in the case of a tungsten core, is a pure SiC, whereas carbon can be included in the mantle if a carbon core is used. Carbon can migrate to the fiber surface which, in any case, is further modified to prevent loss of fiber properties during composite manufacture. The fibers produced by Textron are known as SCS-n [3], where n is the number of microns of surface coating added onto the diameter. The variations of the silicon/carbon content of this layer is closely controlled to permit optimum fiber protection for different matrices so that the SCS-6 fiber is seen as being particularly attractive for titanium–matrix composites. These SiC fibers have very good mechanical properties with a Young's modulus greater than 400 GPa, as shown in Table 2, and can be used at high temperatures around 1000°C, although reactions between the fibers and the matrices they are reinforcing may limit their temperature range in a particular system. Similar SiC fibers, using a tungsten core, are made by DERA in the United Kingdom. These fibers have a diameter of 100 μm.

IV. CARBON FIBERS

Carbon fibers made by carbonizing bamboo fibers were the first filaments used in Edison's incandescent electric lamps but were extremely brittle and rapidly replaced by tungsten wire. Many fibers can be converted into carbon fibers, the basic requirement being that the precursor fiber carbonizes rather than melts when heated [4]. The route adopted in the United States in the 1950s and early 1960s for producing fibers from the next lightest element after boron was to use fibers regenerated from cellulose. This proved a slow process, as the carbon yield is only 24%; however, such fibers are of interest for their thermal conduction properties and are still used in carbon–carbon heat shields and brake pads [5]. The approach taken in Great Britain and Japan was to use polyacrylonitrile (PAN) fibers as precursors for making carbon fibers. An alternative route for making carbon fibers is from pitch obtained either from the residue of oil refining or the coking of coal process.

A. Carbon Fibers from PAN Precursors

This route based on a modified form of acrylic textile fibers proved successful and is the technique used today to make, by far, the greatest number of carbon fibers. Polyacrylonitrile has a carbon yield of 49% and the properties of carbon fibers made by this route depend on the temperature of pyrolysis used so that it is possible to produce a family of carbon fibers with different structures [6]. The strength of carbon fibers reaches a maximum at around 1500°C, whereas the elastic modulus increases continuously with increasing temperature up to around 2800°C. The polyacrylonitrile molecule structure is linear and conversion into carbon fiber requires it to be converted into a three-dimensional structure and the elimination of all atoms except the carbon atoms. This is achieved by first heating the fiber to about 250°C under a tensile load in air, which produces a change in color from white to black and the cross-linking of the structure. Further heating in an inert atmosphere to above 1000°C eliminates most of the elements other than carbon. Above 1500°C, the presence of elements other than carbon is negligible. The structure which results from the pyrolysis of PAN is highly anisotropic, with the basic structural units formed by the carbon atom groups aligned parallel to the fiber axis. There is complete rotational disorder in the radial direction, and the relatively poor stacking of the carbon atoms means that a graphite structure is never achieved. For this reason, it is incorrect to refer to carbon fibers made from PAN as graphite fibers. The structure contains pores which account for the density of the fibers being less than that of fully dense carbon [7]. Table 3 shows some typical properties of carbon fibers; however, there exists a considerable range of properties for these fibers.

B. Carbon Fibers from Pitch-Based Precursors

The high carbon yield of pitch which approaches 90% makes it an attractive and inexpensive source for making carbon fiber precursors. However, cost is increased by the purification processes which are necessary before spinning. The pitch is converted into a mesophase or liquid-crystal solution which is then spun, giving precursor fibers with aligned microstructure [6]. From this point, the carbon fiber fabrication route is the same as that for the PAN-based fibers. However, the response to heating of the pitch-based precursors is not the same. There is no peak in strength at 1500°C, as this is due in the PAN-based fibers to the elimination of nitrogen which is followed by the growth of carbon structural units resulting in a fall in strength. Pitch-based precursors heated to around 2300°C give fibers with Young's moduli as high as those obtained with PAN-based fibers at 2900°C and heating to these higher temperatures gives even greater stiffness. It is therefore more economical to produce high-modulus carbon fibers from pitch than it is from PAN. The properties are due to a less disordered, more graphitic microstructure of the pitch-based fibers, which, however, leads to lower compressive strength and relatively increased cost in producing high-strength carbon fibers.

Carbon fibers of all types need modification of their surfaces, as planes of carbon atoms in the basis structural units of the fiber are generally oriented paral-

Bunsell and Berger

TABLE 3 Typical Characteristics of Carbon Fibers

Fiber type	Manufacturer	Trade mark	Diameter (μm)	Density (g/cm^3)	Strength (GPa)	Strain to failure (%)	Young's modulus (GPa)
Ex PAN							
High strength	Toray	T400H	7	1.80	4.4	1.8	250
High strength	Toray	T1000	5	1.82	7.1	2.4	294
High modulus	Toray	M46J	7	1.84	4.2	1.0	436
High modulus	Toray	M60J	5	1.94	3.92	0.7	588
Ex pitch							
Oil-derived pitch	Nippon Oil	Granoc XN-40	11	2.10	3.7	0.9	390
Oil-derived pitch —high modulus	Nippon Oil	Granoc	11	2.16	3.5	0.5	780
Coal tar pitch	Mitsubishi Chemicals	Dialead K1352U	10	2.12	3.6	0.58	620
Coal tar pitch	Mitsubishi Chemicals	Dialead K13B2U	10	2.16	3.9	0.48	830

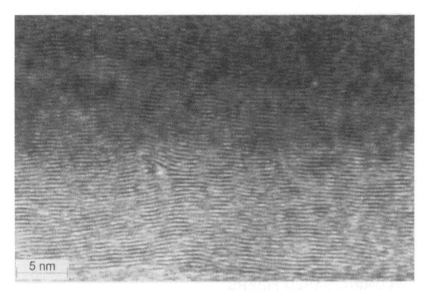

FIG. 2 C_{002} lattice fringe image of a pitch-based fiber.

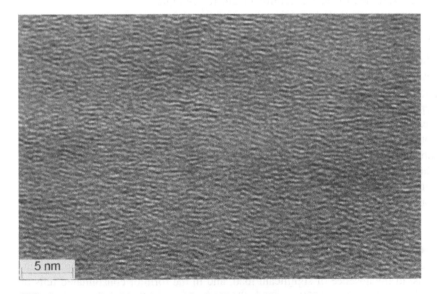

FIG. 3 C_{002} lattice fringe image of a PAN-based fiber.

lel to the surface of the fiber and so present very few opportunities for bonds to be created with the matrix material. Surface treatment takes place usually by anodic oxidation treatment or processes producing a similar oxidation of the fiber surface so as to form available sites for bond formation. This treatment leads to much increased interlaminar shear strengths of the composites which are produced.

Pitch-based carbon fibers are generally less reactive than PAN-based fibers due to their greater microstructural organization. This can be seen by comparing the microstructures shown in Figs. 2 and 3, which show the orientation and organization of a pitch-based carbon fiber and a PAN-based carbon fiber, respectively. In all carbon fibers, the aromatic carbon layers are generally parallel to the fiber surface, which leads to fewer sites being available for forming bonds; however, all carbon fibers suffer from oxidation if heated in air much above 400°C. It is necessary to look elsewhere for reinforcements for use in such oxidative conditions.

V. ALUMINA-BASED FIBERS

Later fibers based on alumina, often combined with silica, were produced from slurries or sol-gels and then sintered. This family of oxide fibers is inherently resistant to oxidation at high temperatures and was originally produced for refractory insulation, then used to reinforce light metal alloys and related fibers are being examined for use as reinforcement for ceramics.

A. Alumina Silica Fibers

The difficulties in producing pure α-alumina fibers, which are the control of porosity and grain growth as well as the brittleness of these fibers, can be overcome by the inclusion of silica in the structure. The microstructures of these fibers depend on the highest temperature the fibers have seen during the ceramization. Very small grains of η-, γ-, or δ-alumina in an amorphous silica continuum are obtained with temperatures below 1000–1100°C. Above this range of temperatures, a rapid growth of α-alumina grains is observed in pure alumina fibers [8]. The introduction of silica allows this transformation to be limited, as it reacts with alumina to form mullite ($\approx 3Al_2O_3 \cdot 2SiO_2$). The presence of mullite at grain boundaries controls the growth of the α-alumina which has not been consumed by the reaction.

The Young's moduli of these fibers are lower compared to that of pure alumina fibers, and such fibers are produced at a lower cost. This, added to easier handling due to their lower stiffness, makes them attractive for thermal insulation applications, in the absence of significant load, and in the form of consolidated felts or bricks up to at least 1500°C. Such fibers are also used to reinforce aluminum alloys in the temperature range 300–350°C. Continuous fibers of this type can be woven due to their lower Young's moduli.

The temperature range at which tensile properties begin to be degraded for fibers containing silica is the same as that for pure alumina fibers. However, higher creep rates are obtained compared to those of pure alumina fibers.

1. The Saffil Fiber

The Saffil fiber [8], which contains 4% silica, is produced by the blow extrusion of partially hydrolyzed solutions of some aluminum salts with a small amount of silica, in which the liquid was extruded through apertures into a high-velocity gas stream. The fiber contains mainly small δ-alumina grains of around 50 nm but also some α-alumina grains of 100 nm. The widest use of the Saffil-type fiber in composites is in the form of a mat which can be shaped to the form desired and then infiltrated with molten metal, usually aluminum alloy. It is the most successful fiber reinforcement for a metal–matrix composite.

For refractory insulation applications, heat treatments of the fiber above 1000°C induce the δ-alumina to progressively change into α-alumina. After 100 h at 1200°C or 1 h at 1400°C, acicular α-alumina grains can be seen on the surface of the fiber and mullite is detected. After 2 h at 1400°C, the transformation is complete and the equilibrium mullite concentration of 13% is established. Shrinkage of the fiber and, hence, the dimensions of bricks are controlled up to at least 1500°C.

2. The Altex Fiber

The Altex fiber is a fiber from Sumitomo Chemicals [9]. The fiber is cylindrical and has a smooth surface. It is produced in two forms which differ in diameter, either 9 or 17 μm. The fiber is obtained by the chemical conversion of a polymeric precursor fiber made from a polyaluminoxane dissolved in an organic solvent to give a viscous product with an alkyl silicate added to provide silica. The precursor is then heated in air to 760°C, a treatment which carbonizes the organic groups to give a ceramic fiber composed of 85% alumina and 15% amorphous silica. The fiber is then heated to 970°C and its microstructure consists of small γ-alumina grains of a few tens of nanometers intimately dispersed in an amorphous silica phase. The fiber has a room-temperature tensile strength of 1.8 GPa and an elastic modulus of 210 GPa. Failure initiation is mainly located at the surface of the fibers.

Subsequent heat treatment produces mullite above 1100°C. At 1400°C, the conversion to mullite is completed and the fiber is composed of 55% mullite and 45% α-alumina by weight.

The presence of silica in the Altex fibers does not reduce their strength at lower temperatures compared with pure alumina fibers; however, a lower activation energy is required for the creep of the fiber. At 1200°C, the continuum of silica allows Newtonian creep and the creep rates are higher than those of Fiber FP, which is composed solely of α-alumina and is described in Section V.2.1.

3. The Nextel 312, 440, and 480 Series of Fibers

The 3M corporation produces a range of ceramic fibers under the general name of Nextel [10]. The Nextel 312, 440, and 480 series of fibers are produced by a sol-gel process. They are composed of 3 mol of alumina for 2 mol of silica, with various amounts of boria to restrict crystal growth. Solvent loss and shrinkage during the drying of the filament produces oval cross sections with the major diameter up to twice the minor diameter, as illustrated in Fig. 4. They are available with average calculated equivalent diameters of 8–9 μm and 10–12 μm and their mechanical properties are reported in Table 4.

The Nextel 312 fiber, first appearing in 1974, is composed of 62 wt% Al_2O_3, 24 wt% SiO_2, and 14 wt% B_2O_3 and appears mainly amorphous from transmission electron microscopic (TEM) observation, although small crystals of aluminum borate have been reported. It has the lowest production cost of the three fibers and is widely used but has a mediocre thermal stability, as boria compounds volatilize from 1000°C, inducing some severe shrinkage above 1200°C. To improve the high-temperature stability in the Nextel 440 and 480 fibers, the amount

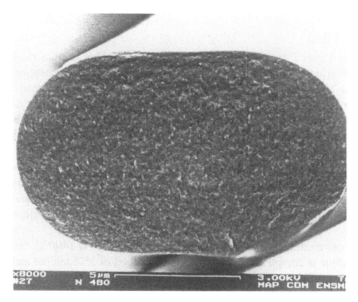

FIG. 4 Appearance of the Nextel 480 fiber. The sol-gel process used produces oval cross section due to shrinkage of the fibers.

TABLE 4 Properties and Compositions of Alumina-Based Fibers

Fiber type	Manufacturer	Trade mark	Composition (wt%)	Diameter (μm)	Density (g/cm³)	Strength (GPa)	Strain to failure (%)	Young's modulus (GPa)
α-Al₂O₃ based fibers	Du Pont de Nemours	FP	99.9% Al₂O₃	20	3.92	1.2	0.29	414
	Du Pont de Nemours	PRD 166	80% Al₂O₃, 20% SiO₂	20	4.2	1.46	0.4	366
	Mitsui Mining	Almax	99.9% Al₂O₃	10	3.6	1.02	0.3	344
	3M	610	99% Al₂O₃, 0.2–0.3% SiO₂ 0.4–0.7% Fe₂O₃	10–12	3.75	1.9	0.5	370
Alumina silica-Based fibers	ICI	Saffil	95% Al₂O₃, 5% SiO₂	1–5	3.2	2	0.67	300
	Sumitomo	Altex	85% Al₂O₃, 15% SiO₂	15	3.2	1.8	0.8	210
	3M	312	62% Al₂O₃, 24% SiO₂, 14% B₂O₃	10–12 or 8–9	2.7	1.7	1.12	152
	3M	440	70% Al₂O₃, 28% SiO₂, 2% B₂O₃	10–12	3.05	2.1	1.11	190
	3M	480	70% Al₂O₃, 28% SiO₂, 2% B₂O₃	10–12	3.05	1.9	0.86	220
	3M	550	73% Al₂O₃, 27% SiO₂	10–12	3.03	2.2	0.98	220
	3M	720	85% Al₂O₃, 15% SiO₂	12	3.4	2.1	0.81	260

of boria has been reduced. These latter fibers have the same compositions: 70% Al_2O_3, 28% SiO_2, and 2% B_2O_3 in weight, but their microstructures are different. Nextel 440 is formed in the main of small γ-alumina in amorphous silica, whereas Nextel 480 is composed of mullite. These differences may be due to different heat treatments of similar initial fibers, the Nextel 440 fiber being heated below the temperature of mulitization. The Nextel 480 is no longer produced commercially.

The evolution of the strengths and creep rates with temperature published by 3M are given in Figs. 5 and 6 [11]. The activation energy for creep of 472 kJ/mol has been obtained for the Nextel 480 fiber, which is close to that found for the Altex fiber.

4. The Nextel 720 Fiber

The Nextel 720 contains the same alumina-to-silica ratio as in the Altex fiber (i.e., around 85 wt% Al_2O_3 and 15 wt% SiO_2 [12]). The fiber, shown in Fig. 7, has a circular cross section and a diameter of 12 μm. The sol-gel route and higher processing temperatures have induced the growth of alumina-rich mullite and α-alumina. Unlike other alumina silica fibers, the Nextel 720 fiber is composed of mosaic grains of ~0.5 μm with wavy contours, consisting of several slightly mutually misoriented grains and elongated grains, as seen in Fig. 8. Its Young's modulus and tensile strength at 25 mm are 260 GPa and 2.1 GPa, respectively. Post-heat-treatment leads to an enrichment of α-alumina in the fiber, as mullite rejects alumina to evolve toward a 3:2 equilibrium composition. Grain growth occurs from 1300°C, and at 1400°C, the wavy interfaces are replaced by straight

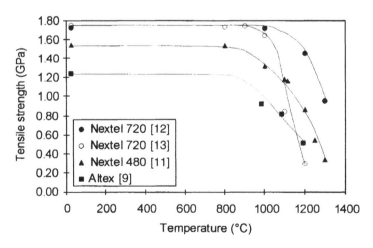

FIG. 5   Temperature dependence of strength for Altex, Nextel 480, and Nextel 720 fibers.

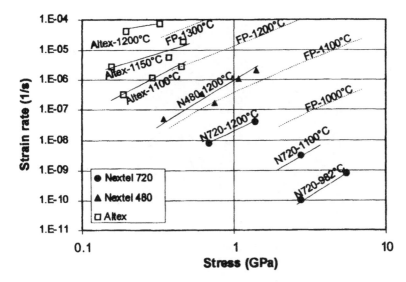

FIG. 6 Evolution of strain rate with applied stress at various temperatures for Altex, Nextel 480, and Nextel 720 fibers. Results obtained with the pure α-alumina FP fiber are given for comparison.

boundaries. A dramatic reduction in creep rate is reported by 3M compared to pure alumina fibers. Dorn plots, shown in Fig. 6 [12], indicate that two orders of magnitude exist between the strain rates of the FP and the Nextel 720 fibers, at 1200°C. This improved creep resistance is attributed to the particular microstructure of the fiber which inhibits the flow of the material. Deléglise et al. [13] have confirmed that the fiber exhibits the lowest creep rates of all commercial small-diameter oxide fibers but have shown that the fiber surface is particularly reactive above 1100°C, which leads to the development of grains as the surface. Slow crack growth initiated by these defects considerably reduces the time to failure during creep as well as the tensile strength of the fiber.

B. α-Alumina Fibers

α-Alumina is the most stable and crystalline form of alumina to which all other phases are converted upon heating above around 1000°C. As we have seen earlier, fibers based on alumina can contain silica, as its presence allows the rapid growth of α-alumina grains to be controlled. However, the presence of silica reduces the Young's modulus of the fiber and reduces their creep strength. High creep resistance implies the production of almost pure α-alumina fibers; however, to obtain a fine and dense microstructure is difficult. The control of grain growth and poros-

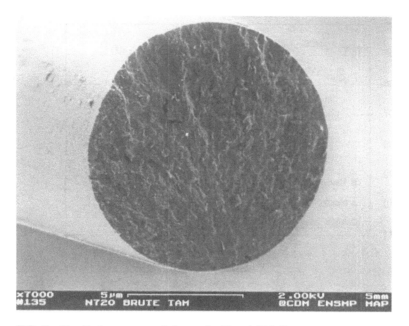

FIG. 7   Tensile fracture morphology of a Nextel 720 fiber.

ity in the production of α-alumina fibers is obtained by using a slurry consisting of α-alumina particles, of strictly controlled granulometry, in an aqueous solution of aluminum salts. The rheology of the slurry is controlled through its water content. The precursor filament then produced by dry spinning is pyrolysed to give an α-alumina fiber.

1.   Fiber FP

The FP fiber, manufactured by Du Pont in 1979, was the first wholly α-alumina fiber to be produced. Its production involved the spinning in air of a slurry composed of an aqueous suspension of Al_2O_3 particles and aluminum salts. The as-obtained fiber was then dried and fired in two steps: the first to control shrinkage and followed, at a higher temperature, by flame firing to obtain a dense microstructure of α-alumina. A final step, involving a brief exposure to a high-temperature flame to produce a fine surface layer of silica, had the effect of improving fiber strength and aiding wettability with metal matrices [14]. It was a continuous fiber with a diameter of 18 μm. This fiber was composed of 99.9% α-alumina and had a density of 3.92 g/cm³ and a polycrystalline microstructure with a grain size of 0.5 μm as seen in Fig. 9, a high Young's modulus of 410 GPa, a tensile strength of 1.55 GPa at 25mm, but a strain to failure of only 0.4%. This brittleness

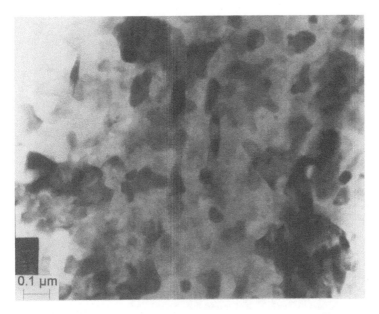

FIG. 8 TEM image showing the microstructure of a Nextel 720 fiber, composed of elongated and mosaic grains.

made it unsuitable for weaving and, although showing initial success as a reinforcement for light alloys, production did not progress beyond the pilot-plant stage and commercial production ceased. Nevertheless, Fiber FP represents an example of an almost pure alumina in filament form and, as such, allows the fundamental mechanisms in this class of fiber to be investigated.

Up to 1000°C, the Fiber FP showed linear macroscopic elastic behavior in tension. Above 1000°C, the fiber was seen to deform plasticity in tension and the mechanical characteristics decreased rapidly, as shown in Fig. 10. At 1300°C, strains in traction increased and could sometimes reach 15%.

Creep was observed from 1000°C. Very little primary creep was reported, but steady-state creep was seen followed by tertiary creep. This small and continuous increase of the strain just before failure indicated an accumulation of damage in the fiber that preceded failure. The strain rates from 1000°C to 1300°C were seen to a function of the square of the applied stress, as seen in Fig. 11, and the activation energy were found to be in the range of 550 to 590 kJ/mol [15,16]. The creep mechanism of Fiber FP has been described as being based on grain-boundary sliding achieved by an intergranular movement of dislocations and accommodated by several interfacial controlled-diffusion mechanisms, involving boundary migration and grain growth. No modification of the granulometry was

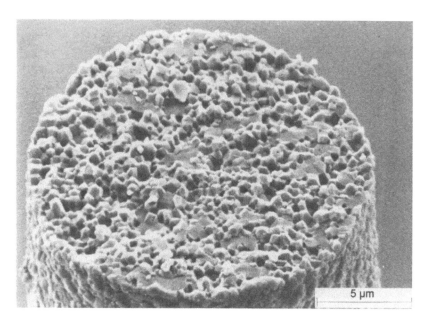

FIG. 9 Fracture morphology of a FP fiber tested at room temperature, showing that the fiber consists of isotropic grains of 0.5 μm.

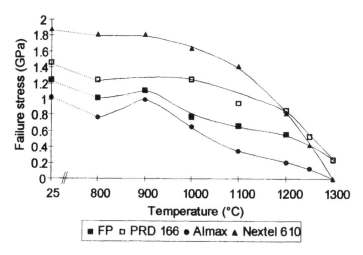

FIG. 10 Temperature dependence of the tensile strengths of Fiber FP, PRD-166, Almax [15], and Nextel 610 [18] fibers.

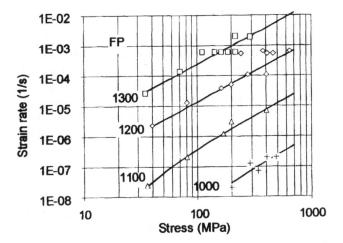

FIG. 11 Evolution of strain rate with applied stress at various temperatures for Fiber FP. (From Ref. 15.)

observed after heat treatment without load at 1300°C for 24 h, but large deformations resulting from tensile and creep tests conducted at 1300°C were observed to induce grain growth. For example, grain growth of 40% was revealed for a fiber which failed with a strain of 30%, with an applied stress equal to 17% of the failure stress. There was no overall preferential direction for the grain growth, but the development of cavities at some triple points was noted due to the pileup of intergranular dislocations at triple points caused by insufficient accommodation of the deformation. The external surfaces of the FP fibers broken in creep at 1300°C after large deformations showed numerous transverse microcracks, which were not observed for smaller strains. The failure of the fiber, at high temperature, occurred after a short period of damage, by the growth of transverse intergranular microcracks from the cavities, the coalescence of which led to a nonflat failure surface.

This fiber was seen to be chemicaly stable at high temperature in air; however, its isotropic fine-grained microstructure led to easy grain sliding and creep, excluding any application as a reinforcement for ceramic structures.

Other manufacturers have modified the production technique to reduce the diameter of the α-alumina fibers that they have produced. This reduction of diameter has an immediate advantage of increasing the flexibility and hence the weaveability of the fibers. Mitsui Mining and 3M Corporation have introduced polycrystalline fibers, the Almax [17] and the Nextel 610 [18] fibers with diameters of 10 μm, half the diameter of Fiber FP.

2. Almax Fiber

An α-alumina fiber, which is still commercially available was produced first in
the early 1990s by Mitsui Mining [17]. It is composed of almost pure α-alumina
and has a diameter of 10 μm. The fiber has a lower density of 3.60 g/cm³ compared to Fiber FP. Like Fiber FP, the Almax fiber consists of one population of
grains of around 0.5 μm; however, the fiber exhibits a large amount of intragranular porosity, as seen in Fig. 12, and is associated with numerous intragranular
dislocations without any periodic arrangement. This indicates rapid grain growth
of α-alumina grains during the fiber fabrication process without elimination of
porosity and internal stresses.

As a consequence, grain growth at 1300°C is activated without an applied
load and reaches 40% after 24 h, unlike that with the other pure α-alumina fibers,
for which grain growth is related to the accommodation of the slip by diffusion.

The fiber exhibits linear elastic behavior at room temperature in tension and
brittle failure. The mechanical properties of the Almax fiber determined by Lavaste et al. [15] for a 25-mm gauge length are as follows: $\sigma_R = 1.7$ GPa, $\varepsilon_R =$
0.5%, and $E = 340$ GPa. The Young's modulus of the Almax fiber is lower than
that of the Fiber FP, because of the greater amount of porosity. The reduction
of the measured failure stress of the Almax compared to those of the FP and the

FIG. 12   Appearance of the Almax fiber. The fiber is composed of grains of 0.5 μm with
a significant amount of intragranular porosity.

more pronounced intragranular failure mode for this fiber compared to the FP fiber show a weakening of the grains by the intragranular porosity.

The Almax fibers exhibit linear macroscopic elastic behavior up to 1000°C. Above 1000°C, the mechanical characteristics decrease rapidly (Fig. 10), with a more severe drop than for Fiber FP. Tensile failure of the Almax fiber at 1250°C has revealed isotropic grain growth up to 55%. No extended regions of damage could be observed, as seen on the Fiber FP surface

Creep occurs from 1000°C for the Almax fibers and the fibers show a lower resistance to creep than Fiber FP fibers, as shown in Fig. 13. Diffusion and grain-boundary sliding are facilitated by the growth of intergranular porosity and much higher strain rates are obtained. This intergranular porosity which appears in the fiber during creep, may have been created by the interception of intragranular pores by the boundaries of the growing grains. This intergranular pore growth induces failure rapidly.

3. Nextel 610

A continuous α-alumina fiber, with a diameter of 10 μm, was introduced by the 3M Corporation in the early 1990s with the trade name Nextel 610 fiber [18]. It

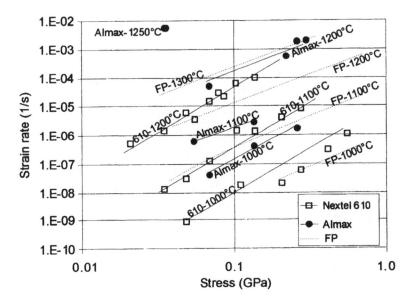

FIG. 13 Evolution of strain rate with applied stress at various temperatures for Almax [15] and Nextel 610 [18] α-alumina fibers. Results obtained with the pure dense α-alumina FP fiber are given for comparison.

is shown in Fig. 14 and is composed of around 99% α-alumina, although a more detailed chemical analysis gives 1.15% total impurities, including 0.67% Fe_2O_3 used as a nucleating agent and 0.35% SiO_2 as the grain growth inhibitor. It is believed that the silica, which is introduced, does not form a second phase at grain boundaries, although the suggestion of a very thin second phase separating most of the grains has been observed by transmission electron microscopy. The fiber is polycrystalline with a grain size of 0.1 μm, five times smaller than in Fiber FP. As shown in Table 4, the strength announced by 3M is 2.4 GPa, for a gauge length of 106 mm, which is twice the tensile strength measured on Fiber FP for a gauge length of 150 mm, and the elastic modulus is 380 GPa.

Creep occurred from 900°C and strain rates are two to six times larger than those of Fiber FP (Fig. 13) due to a finer granulometry and possibly to the chemistry of its grain boundaries. A stress exponent of approximately 3 is found between 1000°C and 1200°C, with an apparent activation energy of 660 kJ/mol [18]. Failure frequently occurs via the coalescence of cavities into large cracks over the whole gauge length and the failure surfaces are rougher compared to those obtained at room temperature, in the same manner as was seen with Fiber FP. Local cavitations and necking around heterogeneities sometimes induce failure.

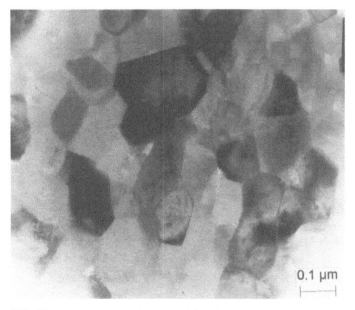

FIG. 14   Microstructure of the Nextel 610 fiber composed of grains of 0.1 μm on average.

4. PRD-166 Fiber

Du Pont synthesized the PRD-166 fiber [19] in which 20 wt% of partially stabilized zirconia was added to increase the elongation to failure of the fiber. The intention was to produce a fiber which, compared to Fiber FP, was easier to weave. The PRD-166 fiber had a diameter of 18 μm and a density of 4.2 g/cm³. The dispersion of zirconia intergranular particles of 0.15 μm limits grain growth of the alumina grains, which have a mean diameter of 0.3 μm, as seen in Fig. 15, instead of 0.5 μm for Fiber FP for a similar initial alumina powder granulometry. These particles underwent a martinsitic reaction in the vicinity of the crack tips, resulting in the partial closure of cracks and an increase of the fiber strength to $\sigma_R = 1.8$ GPa at a gauge length of 25 mm [15]. The resulting stiffness of the reinforced alumina is lower than that of Fiber FP, $E = 344$ GPa, due to the lower Young's modulus of zirconia compared to that of alumina. The increase in strain to failure was not sufficient to allow weaving with the PRD-166 fiber, and production of the PRD-166 fiber did not progress beyond the pilot stage; however, the study of this fiber permits a greater understanding of the mechanisms of toughening and the enhancement of creep behavior of alumina fibers.

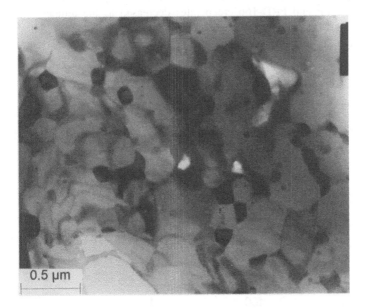

0.5 μm

FIG. 15 Microstructure of PRD-166 fiber. Brighter alumina grains of 0.3 μm on average are surrounded by darker zirconia grains of 0.1 μm. Smaller spherical zirconia grains can be observed in the alumina grains.

At high temperatures, the introduction of dispersed particles of zirconia at triple points of grain boundaries limited the mobility of intergranular dislocations from one grain boundary to another. The PRD-166 fibers conserved their elastic properties and creep strength up to 1100°C (Fig. 10), which is 100°C above Fiber FP. Greater times to failure and lower creep strain rates were obtained for the PRD-166 compared to the FP, as seen in Fig. 16 [15]. At 1000°C and 1100°C, the strain rates were seen to be a function of $(\sigma - \sigma_0)^2$, indicating that a significant threshold stress σ_0 needed to be reached to permit creep. The threshold stress was equal to 180 MPa at 1000°C and 90 MPa at 1100°C and could be neglected at higher temperatures. A higher activation energy for creep, $Q = 600$ kJ/mol, was required compared to Fiber FP. From 1200°C, a regrouping of zirconia grains into chains became possible, thus decreasing the surface energy of the grains. This is characteristic of a two-phase material in which the diffusivity of one species, here Zr^{4+}, through the boundary is higher than the other. The thermally activated intergranular diffusion and motion of dislocations from one boundary to another were facilitated, and grain growth was no longer impeded, resulting in a higher growth rate at 1300°C than with the Fiber FP. For example a 70% increase in average grain size was noticed for both alumina and zirconia for a fiber failed in tension at 1300°C at a strain of 7%. As a consequence, the differences between the FP and PRD-166 creep rates decreased as the temperature increased, and at 1300°C, the strain rates were almost the same.

A comparison of the duration of the nonlinear creep regions and of the external surfaces of the fibers after straining showed that damage in Fiber FP rapidly

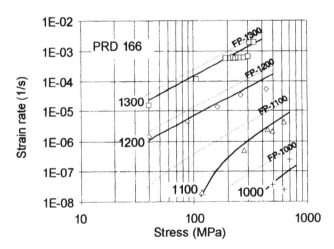

FIG. 16 Evolution of strain rate with applied stress at various temperature for the PRD-166 fiber. (From Ref. 15.)

induced failure, whereas the PRD-166 fiber resisted higher levels of damage. Although the transformation at high temperatures of the tetragonal particles to the monoclinic form was inhibited, they still toughened the fiber by deflecting cracks. However, the development of dispersed microcracks progressively decreased the load-supporting cross section, resulting in an increase in true stress at constant applied load and an increase of the strain rate.

Existing polycrystalline fibers based on alumina have been seen to be resistant to oxidation but lose their mechanical properties above 1200°C due to microstructural modifications and enhanced facility of grain motion.

VI. SiC-BASED FIBERS

A third group of fibers, based on silicon carbide, have been developed through the pyrolysis of organo-silicon precursor filaments [20] in an analogous fashion to the technique used to make the most successful carbon fibers, which are produced by carbonizing precursors of polyacrylonitrile filaments. These silicon-carbide-based fibers first allowed ceramic–matrix composites to be developed and are, at present, the most widely used reinforcement for this type of composite. For these applications, the fibers are often made with a carbon-rich surface which is useful in controlling the interface in composites. An alternative surface coating could be boron nitride. Recently, complex surface coatings of SiC and pyrolytic carbon have been prepared to control interfacial debonding [21].

A. NL-200 Series of Nicalon Fibers

The work of Yajima and his colleagues in Japan was first published in the mid-1970s and gave rise to the first and so far the most successful fine ceramic fibers [22]. This fiber is produced by Nippon Carbon under the name of Nicalon. The most widely used fiber in this family at present is the Nicalon NL-200. The manufacture of Nicalon fibers involves the production of polycarbosilane (PCS) precursor fibers which consist of cycles of six atoms arranged in a similar manner to the diamond structure of β-SiC. The molecular weight of this polycarbosilane is low, around 1500, which makes drawing of the fiber difficult. In addition, methyl groups ($-CH_3$) in the polymer are not included in the $Si-C-Si$ chain, so that during pyrolysis, the hydrogen is driven off, leaving a residue of free carbon. The production of the NL-200 fiber involves subjecting the precursor fibers to heating in air at about 300°C to produce cross-linking of the structure. This oxidation makes the fiber infusible but has the drawback of introducing oxygen into the structure which remains after pyrolysis. The ceramic fiber is obtained by a slow increase in temperature in an inert atmosphere up to 1200°C and has a glassy appearance when observed in scanning electron microscopy (SEM), as shown in Fig. 17. The fiber, however, contains a majority of β-SiC,

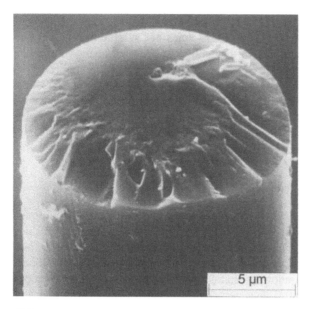

FIG. 17   Appearance of a Nicalon NL-200 fiber.

of around 2 nm, but also significant amounts of free carbon of less than 1 nm
and excess silicon combined with oxygen and carbon as an intergranular phase
[23]. The strengths and Young's moduli of Nicalon fibers tested in air or an inert
atmosphere show little change up to 1000°C, as can be seen in Fig. 18. Above
this temperature, both of these properties show a slight decrease up to 1400°C.
Between 1400 and 1500°C, the intergranular phase begin to decompose, carbon
and silicon monoxides are evacuated, and a rapid grain growth of the silicon
carbide grains is observed. The density of the fiber decreases rapidly and the
tensile properties exhibit a dramatic fall. When a load is applied to the fibers, it
is found that a creep threshold stress exists above which creep occurs. The fiber
is seen to creep above 1000°C and no stress enhanced grain growth is observed
after deformation [24]. Creep is due to the presence of the oxygen-rich intergranular phase. The properties and composition of these fibers are shown in Table 5.

B.   Tyranno LOX-M

The Tyranno LOX-M fiber produced by Ube Industries is made with a precursor
similar to that used for the Nicalon fibers, but with the addition of titanium.
The structure of this polytitanocarbosilane (PTC) precursor is complex due to
condensation of Si—H bonds and cross-linking of titanium alkoxide occurring

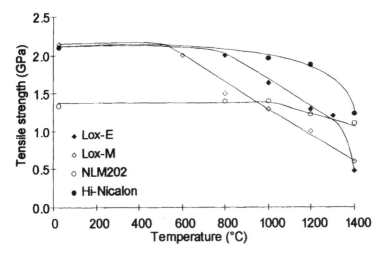

FIG. 18 Evolution of tensile strengths of the NL-200 and Hi-Nicalon fibers and Tyranno LOX-M and LOX-E fibers with temperature. (From Ref. 24.)

simultaneously [25]. The precursor fiber is cross-linked by heating in air at about 180°C and then converted into a ceramic fiber by treatment in nitrogen above 1000°C. The resulting fiber contains a small amount of titanium, around 2% by weight, which is said to inhibit crystallization and offer better resistance to oxidation of the carbon by the formation of Ti—C bonds. Titanium carbide grains are seen from 1200°C. The Tyranno LOX-M fiber contains 13 wt% of oxygen, which limits its use over 1000°C because of the same processes seen in the Nicalon NL-200 fiber [24]. The Tyranno LOX-M fiber has a microstructure similar to that of the NL-200 and can be produced with two diameters of 8.5 μm and 12 μm. The Lox-M fibers have been successfully used for the formation of composite material, without the infiltration of a matrix material and with a high fiber volume fraction, under the name of Tyranno Hex. Bundles of fibers, which have been preoxidized to give them a thin surface layer of silica, are hot pressed, leading to a dense hexagonal packing of the fibers, the cavities being filled by silica and TiC particles. The strength of Tyranno Hex measured in bending tests has been reported to be stable up 1400°C in air [26].

C. Hi-Nicalon and Tyranno LOX-E Fibers

A later generation of Nicalon and Tyranno fibers has been produced by cross-linking the precursors by electron irradiation to avoid the introduction, at this stage, of oxygen [27]. These fibers are known as Hi-Nicalon, which contains 0.5

TABLE 5 Properties and Compositions of Silicon-Based Fibers

Fiber type	Manufacturer	Trade mark	Composition (wt%)	Diameter (μm)	Density (g/cm³)	Strength (GPa)	Strain to failure (%)	Young's modulus (GPa)
Si—C based	Nippon Carbide	Nicalon, NLM 202	56.6% Si, 31.7% C, 11.7% O	14	2.55	2.0	1.05	190
	Nippon Carbide	Hi-Nicalon	62.4% Si, 37.1% C, 0.5% O	14	2.74	2.6	1.0	263
	Ube Chemical	Tyranno Lox-M	54.0% Si, 31.6% C, 12.4% O, 2.0% Ti	8.5	2.37	2.5	1.4	180
	Ube Chemical	Tyranno Lox-E	54.8% Si, 37.5% C, 5.8% O, 1.9% Ti	11	2.39	2.9	1.45	199
Si—N	Tonen	Tonen	58.6% Si, 38.2% N, 2.7% O, 0.5% C		2.5	2.5	1	250
Si—N—C based	Dow Corning	HPZ	59% Si, 28% N, 10% C, 3% O	10–12	2.3–2.5	1.7–2.1	~1	180–230

wt% oxygen and Tyranno LOX-E which contains approximately 5 wt% oxygen. The higher value of oxygen in the LOX-E fiber is due to the introduction of titanium alkoxides for the fabrication of the PTC. The decrease in oxygen content in the Hi-Nicalon compared to the NL-200 fibers has resulted in an increase in the size of the SiC grains and a better organization of the free carbon. The mean size of the SiC grains is 5 nm, but crystallites presenting stacking faults can reach 20 nm. Carbon aggregates appear by the stacking of four distorted layers over a length of 2 nm on average. Figure 19 shows that a significant part of the SiC is not perfectly crystallized and surrounds the ovoid β-SiC grains [24]. Further heat treatment of the fibers at 1450°C induces the SiC grain to grow up to a mean size of 20 nm, to develop facets, and to be in contact with adjacent SiC grains [28]. Turbostratic carbon has been seen to grow preferentially parallel to some of these facets and could, in some cases, form cages around SiC grains, limiting their growth. Significant improvements in the creep resistance are found for the Hi-Nicalon fiber compared to the NL-200 fiber, which can further be enhanced by a heat treatment so as to increase its crystalinity. The LOX-E fiber has a microstructure and creep properties comparable to those of the LOX-M fiber [24].

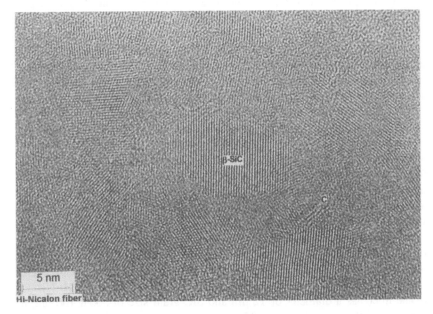

FIG. 19 Nanostructure of the Hi-Nicalon fiber revealed by high-resolution transmission electron microscopy. The fiber is composed of crystalline β-SiC and turbostratic carbon embedded in an imperfectly organized SiC phase.

The size of the SiC grains of the LOX-E fiber only slightly changes after heat treatment at temperatures below 1500°C. Despite the electron curing process, the use of a PTC does not allow a sufficient reduction of the oxygen in the intergranular phase of the ceramic fiber to the extent seen in the Hi-Nicalon, so that grain growth is impeded and creep enhanced. A more recent polymer, polyzirconocarbosilane (PZT), has allowed the titanium to be replaced by zirconium and the oxygen content to be reduced [29]. The resulting fibers, known as Tyranno ZE and which contains 2 wt% of oxygen, show increased high-temperature creep and chemical stability and resistance to corrosive environments.

D. Near-Stoichiometric Fibers

Efforts to reduce the oxygen content by processing in inert atmospheres and cross-linking by radiation have produced fibers with very low oxygen contents. These fibers are not, however, stoichiometric, as they contain significant amount of excess free carbon affecting oxidative stability and creep resistance. Near-stoichiometric SiC fibers from polymer precursors are being developed by the principal fiber producers at a laboratory scale in the United States and Japan.

Dow Corning has produced stoichiometric SiC fibers from a PCS containing a small amount of titanium [30]. The precursor fibers are cured by oxidation and doped with boron. In this way, degradation of the oxicarbide phase at high temperature is controlled, and catastrophic grain growth and associated porosity as occurred with the previous oxygen-rich fibers is avoided. The precursor fiber can then be sintered at high temperature ($\sim$1600°C) so that the excess carbon and oxygen are lost as volatile species to yield a polycrystalline, near-stoichiometric SiC fiber called SYLRAMIC fiber. Such a fiber has a diameter of 10 μm and SiC grain sizes ranging from 0.1 to 0.5 μm with smaller grains of TiB_2 and B_4C (Fig. 20). A carbon- and oxygen-rich surface and a BN-like (boron nitride) component in the near-surface region have been identified. The tensile strength measured with a 25-mm gauge length reaches 3.4 GPa. The elastic modulus of the fiber of 386 GPa and its density of $\sim$3.1 g/cm^3 are slightly less than the values of fully dense SiC. The fiber retains 50% of its initial strength after a heat treatment in dry air at 1200°C lasting 100 h. Characterization of the fiber in creep shows a considerably reduced deformation rate when compared to the Hi-Nicalon fiber.

Nippon Carbon has obtained a near-stoichiometric fiber, the Hi-Nicalon S, from a PCS cured by electron irradiation and pyrolysed by a "modified Hi-Nicalon process" in another closely controlled atmosphere [31]. As a result, the excess carbon is reduced from C/Si = 1.39 for the Hi-Nicalon to 1.05 for the Hi-Nicalon S. The fiber has a diameter of 12 μm and SiC grain size of about 100 nm, its tensile strength measured with a 25-mm gauge length reaches 2.6 GPa, and its elastic modulus 420 GPa for a density of 3.1 g/cm^3. Relaxation tests clearly show an improvement in creep resistance compared to the Hi-Nicalon fiber. After 10

FIG. 20 Microstructure of the SYLRAMIC fiber composed of SiC grains of 100–200 nm and of smaller particles of TiB_2 and B_4C.

h at 1400°C in dry air, the fiber retains 70% of its strength and is covered by an oxide layer of 0.3–0.4 μm without the strength-reducing CO oxide bubbles found in the carbon-rich Hi-Nicalon fiber.

Ube Industries has developed a near-stoichiometric fiber made from a poly-aluminocarbosilane [32]. The precursor fiber is cured by oxidation, pyrolysed up to 1700°C to allow the outgassing of CO, and sintered at a temperature above 1800°C. The fiber has a diameter of 10 μm and SiC grain sizes of about 200 nm, as shown in Fig. 21. Aluminum has been added as a sintering aid and, from the results of the manufacturer, gives a better corrosion resistance than TiB_2 which is found in the SYLRAMIC fiber. The room-temperature properties announced are 3 GPa for the tensile strength and 330 GPa for the Young's modulus, and the creep resistance has been considerably enhanced compared to other nonstoichiometric fibers.

The emerging generation of stoichiometric SiC fibers represents a solution to the instability of earlier fibers; however, the accompanying increase in Young's modulus and a slight loss in strength due to larger grain sizes lead to fibers which become more difficult to handle and convert into structures. This difficulty may be overcome by transforming partially converted Si—C—O fibers into the wo-

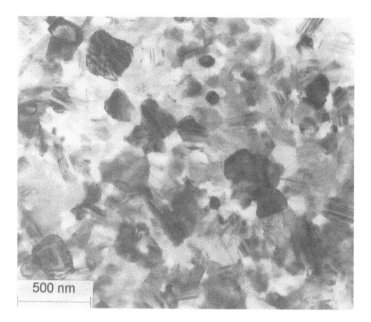

FIG. 21 Microstructure of the Tyranno SA fiber compsed of SiC grains of 200 nm.

ven or other form of fiber arrangement followed by pyrolysis and sintering to convert them into a stoichiometric dense form. The fiber structure could then be infiltrated to form the matrix, giving an optimized ceramic–matrix composite. However, even stoichiometric silicon carbide fibers will suffer from oxidation from 1200°C, resulting in the formation of a silica surface layer which would modify the properties of the SiC–matrix interface in the composites used in oxidizing conditions. For this reason, this family of fibers are likely to be limited to a maximum temperature of 1400°C.

VII. Si—C—N-BASED FIBERS

A. Si—C—N—O Fibers

A group of ceramic fibers which have been studied but which have not given rise to commercial fibers were based on a silicon carbide–silicon nitride system and they can be produced from polycarbosilazane precursors. Various routes for the manufacture of these fibers are possible. The fibers can be produced by the pyrolysis of polysilazanes prepared from tri(N-methylamino) methylsilane polymerized by heating to 520°C for several hours and then heated to between 1200°C and 1500°C for 2 h. During pyrolysis, a weight loss of approximately 35% is reported to occur and, finally, the fibers which are produced are described as being shiny black in appearance.

This type of fiber was initially studied for the dielectric properties which is possible to obtain with this class of material. The electrical resistivity of silicon carbide–silicon nitride is about 1000 times greater than that obtained for graphite. Relatively high mechanical properties have been reported for these fibers with moduli around 200 GPa and breaking strengths of about 1 GPa.

B. Silicon Nitride Fibers

The development of a continuous, 10-μm-diameter, silicon nitride fiber by Tonen has been reported, although it is not generally commercially available. The fiber is said to be amorphous and to retain 90% of its room-temperature strength after exposure in air for 12 h at 1300°C and 70% at 1400°C. After 1 h at 1500°C, strength retention is reported as being 55%. The average room-temperature strength is given by the manufacturer as 2.5 GPa. This fiber is said to retain its amorphous structure to 1400°C, but in the presence of nitrogen, Si_3N_4 crystallizes at the surface. The presence of these crystals causes the fiber to lose its strength.

C. Si—B—(N,C) Fibers

It is reported that a new amorphous single-phase (carbo-)nitride ceramic has been developed by Bayer from a polymer precursor which can be spun and converted into ceramic fibers [33]. The ceramic is reported to be stable in an oxidizing atmosphere up to 1600°C due to the formation of a protective double layer composed of an outer layer of SiO_2 covering a layer of BN(O). This inner layer is said to prevent the penetration of oxygen to the core of the fiber. Initial data from the manufacturer indicate that the fibers possess promising mechanical properties which seem to be preserved to very high temperatures. The manufacturer has indicated that the spinning process used to make this fiber is being optimized.

VIII. CONTINUOUS MONOCRYSTALLINE FILAMENTS

Continuous monocrystalline filaments have been developed by the Saphikon company in the United States. These filaments are grown from molten alumina and, as a consequence, are produced at a slow rate and high cost and with large diameters usually in excess of 100 μm [34]. The near-stoichiometric composition of these fibers with an absence of grain boundaries ensures that they should be able to better withstand high temperatures above 1600°C. Careful orientation of the seed crystal enables the crystalline orientation to be controlled so that creep resistance can be optimized. Published data on the strength of Saphikon fibers as a function of temperature reveals that strength variation is not a single function of temperature. The observed fall in strength around 300°C, which is then fol-

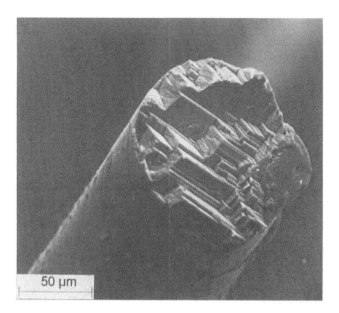

FIG. 22 A typical fracture morphology of the α-alumina Saphikon fiber, revealing the cleavage planes. Failure has been initiated at a process defect.

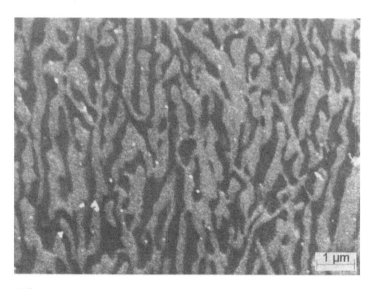

FIG. 23 SEM image of a longitudinal cross section of the eutectic YAG–alumina Saphikon fiber revealing its elongated two-phase microstructure (darkest phase = alumina, brightest phase = YAG).

lowed by an increase in strength around 500°C, could be due to stress corrosion followed by crack blunting. These fibers are not without defects and characteristic bubbles can be seen in the fibers, most probably due to convection during fiber growth at the meniscus point between the solid and the melt. A typical fracture morphology of the α-alumina Saphikon fiber is shown in Fig. 22.

The same manufacturing processes have been employed to produce an eutectic fiber consisting of interpenetrating phases of α-alumina and YAG [35]. The structure depends on the conditions of manufacture, in particular the drawing speed, but can be lamellar and oriented parallel to the fiber axis. Figure 23 is a longitudinal cross section of the fiber, as revealed by SEM, showing the elongated YAG and alumina phases. This fiber does not show the same fall in strength seen with the single-phase alumina fiber. However, such fibers are seen to relax from 1100°C but do not have as strong a dependence on temperature as the polycrystalline oxide fibers.

IX. WHISKERS

Whiskers are monocrystals in the form of filaments. The potential of whiskers as reinforcements has been discussed for many years, as their small diameters, usually between 0.5 and 1.5 μm, means that they contain very few defects and must possess extremely high strengths, perhaps up to the theoretical strength for matter, which is approximately one-tenth of its Young's modulus. In addition, their aspect ratios of length to diameter can be considerable, as they can be produced with lengths between 20 μm and, it is claimed, several centimeters [36]. A high aspect ratio is just what is required to achieve reinforcement. Considerable difficulties have to be overcome if whiskers are to be used as reinforcements, however. They are extremely small, so that plastic bag containing whiskers seems to contain dust. This means that alignment of the whiskers in a matrix is very difficult. There are potential uses for whiskers combined with more conventional fibers so as to provide some reinforcement of the matrix in the transverse direction. Their fineness is also another handicap in their exploitation, as 1 μm is just the size to block the alveolar structure of the lungs. For this reason above all, whiskers remain an intriguing possibility as reinforcements but one which is little exploited.

X. CONCLUSION

Inorganic fibers are both the most common reinforcements for composite materials as well as, some of them, being among the most exotic. Glass fibers represent around 99% of fiber reinforcements. They are made by the extrusion of molten glass through the holes of spinnerettes so as to be formed into virtually continuous filaments with diameters of the order of 10 μm. Glass fibers must be coated as soon as they leave the spinnerette so as to protect them against environmental attack and mechanical damage and to ensure that a good bond is made with the matrix material. Unlike most of the other fibers which have been discussed, glass fibers are amorphous. Carbon fibers form a large family of fine reinforcements with diameters usually considerably less than 10 μm and with a very wide range of properties. They are made by the pyrolysis of organic precursor fibers, and during their fabrication, great care is taken not to lose the alignment which exists in the microstructure of the drawn organic fibers. This alignment and the perfection of the microstructure, which is controlled through the processing temperature, determine their ultimate mechanical properties. Carbon fibers are surface treated through oxidation to create sites for chemical bonds to be formed with the matrix material. The main interest today in fibers made by chemical vapor deposition onto a tungsten or carbon core is for the reinforcement of titanium. Such fibers have been made with boron and silicon carbide and have diameters usually greater than 100 μm. Both boron and silicon carbide fibers have to be surface treated by the deposition of boron carbide or silicon carbide. In the case of the most successful SiC fibers, the coating is one in which the ratio of the silicon and carbon varies so as to provide the protection necessary during metal–matrix composite manufacture.

A family exists of fibers, with diameters of less than 20 μm, based on alumina, often associated with silica. Alumina exists in several phases, but all are converted into the α form at high enough temperatures. The silica is used to retard this conversion and slow grain growth. The fibers have been produced both as refractory insulation and reinforcements for light alloys, but they are also of potential interest because of their chemical inertness in oxidizing atmospheres as reinforcements for ceramic matrices. At high temperatures, the polycrystalline alumina fibers creep due to grain sliding. Much lower creep rates are observed in oxide fibers composed of two distinct phases and it is thought that the elongated form of one of the phases contributes to this creep resistance.

The development of fibers based on silicon carbide made from organo-silicon precursors with diameters less than 15 μm have allowed ceramic–matrix composites to be produced. The silicon carbide phase is in the form of grains with dimensions of several nanometers. They show creep rates at high temperatures which depend on the presence of an intergranular phase existing as a consequence of the manufacturing process. The control of this phase has allowed the fibers to be optimized, and as their composition approaches stoichiometry, their high-temperature properties improve. In such near-stoichiometric fibers, the SiC grains are of the order of 100 nm. The role of these fibers in a ceramic–matrix is to hinder crack propagation in the matrix material. In order to do this, the fibers are usually coated with pyrolytic carbon to prevent the formation of a too strong an interfacial bond.

The control of the fibers' surfaces is therefore of paramount importance in obtaining good composite properties. The desired result in the case of matrices which can undergo plastic deformation, which are the resins and metals, is to obtain an efficient load transfer between the reinforcements and the matrix so that a good bond is required. In contrast, the brittle matrix composites comprising ceramic fibers in a ceramic–matrix need a poor bond which can act as a mechanical fuse to hinder crack propagation. In this latter case, which concerns composites to be used at high temperatures, the development of second phases, most often silica for the SiC-based fibers and other interactions between oxide fibers and oxide matrices, at the surface can weaken the fibers as well as forming a too strong adherence, which reduces the toughness of the composite. The control of these interfacial reactions is, at present, of great research interest.

Inorganic fibers find uses over a very wide range of applications and are set to continue to play a dominant role as reinforcements for composite materials.

REFERENCES

1. R. W. Moncrieff. *Man-made Fibers*, 6th ed., Butterworth, Guildford, 1979.
2. P. K. Gupta, in *Fiber Reinforcements for Composite Materials* (A. R. Bunsell, ed.), Elsevier, Amsterdam, 1988, pp. 19–71.
3. F. E. Wawner, in *Fiber Reinforcements for Composite Materials* (A. R. Bunsell, ed.), Elsevier, Amsterdam, 1988, pp. 371–425.
4. M. Sitting. *Carbon and Graphite Fibers: Manufacture and Applications*, Noyes Dates Co., 1980.
5. A. T. Kaverov, M. E. Kazakov, and V. Ya. Varshavsky, in *Fiber Science and Technology* (V. I. Kostikov, ed.), Chapman & Hall, London, 1995, pp. 231–358.
6. E. Fitzer and M. Heine, in *Fiber Reinforcements for Composite Materials* (A. R. Bunsell, ed.), Elsevier, Amsterdam, 1988, pp. 73–148.
7. A. Oberlin and M. Guigon, in *Fiber Reinforcements for Composite Materials* (A. R. Bunsell, ed.), Elsevier, Amsterdam, 1988, pp. 149–210.
8. J. D. Birchall. Trans. J. Br. Ceram. Soc. 82:143 (1983).
9. Y. Abe, S. Horikiri, K. Fujimura, and E. Ichiki, in *Proceedings of the 4th International Conference on Composite Materials, Tokyo, 1982* (T. Hayashi, K. Kawata, and S. Umekawa, eds.), The Japan Soc. for Composite Materials, Tokyo, 1982, pp. 1427–1434.
10. T. L. Tompkins. *Ceramic Industry*, Business News Publishing Company, 1995.
11. D. D. Johnson, A. R. Holtz, and M. F. Grether. Ceram. Eng. Sci. Proc. 8:744 (1987).
12. D. M. Wilson, S. L. Lieder, and D. C. Lueneburg. Ceram. Eng. Sci. Proc. 16:1005 (1995).
13. F. Deléglise, M. H. Berger, and A. R. Bunsell, in *Proc. 8th European Conference on Composite Materials, Naples, 1998* (C. Visconti, ed.), Woodhead Pub. Ltd., Cambridge, 1998, pp. 175–182.
14. A. K. Dhingra. Phil. Trans. R. Soc. Lond. A 294:411 (1980).
15. V. Lavaste, M. H. Berger, A. R. Bunsell, and J. Besson. J. Mater. Sci. 30:4215 (1995).

16. D. J. Pysher and R. E. Tressler. J. Mater. Sci. *27*:423 (1992).
17. Y. Saitow, K. Iwanaga, S. Itou, et al., Proc. of the 37th International SAMPE Symposium and Exhibition, 1992, pp. 808–819.
18. D. M. Wilson, D. C. Lueneburg, and S. L. Lieder. Ceram. Eng. Sci. Proc. *14*:609 (1993).
19. J. C. Romine. Ceram. Eng. Sci. Proc. *8*:755 (1987).
20. J. Lipowitz. Ceram. Bull. *70*:1888 (1991).
21. R. R. Naslain. Composites Pt. A *29A*:1145 (1998).
22. S. Yajima, J. Hayashi, and M. Omori. Chem. Lett. *9*:931 (1975).
23. P. Le Coustumer, M. Monthioux, and A. Oberlin, J. Eur. Ceram. Soc. *11*:95 (1993).
24. N. Hochet, M. H. Berger, and A. R. Bunsell. J. Microsc. *185*:243 (1997).
25. T. Yamamura, T. Ishikawa, M. Shibuya, T. Hisayuky, and K. Okamura. J. Mater. Sci. *23*:2589 (1988).
26. T. Yamamura, T. Ishikawa, M. Sato, M. Shibuya, H. Ohtsubo, T. Nagasawa, and K. Okamura. Ceram. Eng. Sci. Proc. *11*(9–10):1648 (1990).
27. T. Seguchi, M. Sugimoto, and K. Okamura, in *High Temperature Ceramic Matrix Composites I* (R. Naslain et al., eds.), Woodhead Publishing Ltd., Cambridge, 1993, p. 51.
28. M. H. Berger, N. Hochet, and A. R. Bunsell. Ceram. Eng. Sci. Proc. *19*:39–46 (1998).
29. K. Kumagawa, Y. Yamaoka, M. Shibuya, and T. Yamamura. Ceram. Eng. Sci. Proc. *18*:113–118 (1997).
30. J. Lipowitz, J. A. Rabe, A. Zangvil, and Y. Xu. Ceram. Eng. Sci. Proc. *18*:147–157 (1997).
31. H. Ichikawa, K. Okamura, and T. Seguchi, in *Proceedings of the Second International Conference on High Temperature Ceramic Matrix*, Ceramic (A. G. Evans and R. Naslain, eds.), The American Ceramic Society, Santa Barbara, CA, 1995, pp. 65–74.
32. T. Ishikawa, S. Kajii, T. Hisayuki, K. Matsunaga, T. Hogami, and Y. Kohtoku. Key Eng. Mater. *164–165*:15 (1999).
33. H. P. Baldus, G. Passing, D. Sporn, and A. Thierauf. Ceram. Trans. *58*:75 (1995).
34. J. T. A. Pollock. J. Mater. Sci. *7*:631 (1972).
35. T. A. Parthasarathy, T. Mah, and L. E. Matson. Ceram. Eng. Sci. Proc. *11*(9–10):1628 (1990).
36. W. E. Hollar and J. J. Kim, Jr. Ceram. Eng. Sci. Proc. *12*(7–8):979 (1991).

8

Surface Modification of Textiles by Plasma Treatments

CEZAR-DORU RADU   Department of Textile Finishing, Technical University of Iasi, Iasi, Romania

PAUL KIEKENS and JO VERSCHUREN   Department of Textiles, University of Gent, Gent, Belgium

I. INTRODUCTION

The idea of the treatment of textiles with plasma is a few decennia old. There is no consensus about who was "the first," but it is clear that the treatment of textiles is historically linked to the plasma treatment of polymers in general.

As one of the most promising alternatives in many fields, the importance of plasma treatments results from the exceptional advantages it offers. It does have specific action only at the surface, keeping the bulk properties unaffected. The future of plasma is closely linked to the fact that this technique gives the treated surface some properties that cannot be obtained by conventional techniques, and this is without the need to use water as a reaction medium. At the level of textiles, this means changing an almost inert surface into a reactive one, and in this way, it becomes a surface engineering tool. The transfer of research results into the technological field would lead to nonpolluting and very promising operating conditions. In the prospect of chemical finishing using plasma, two main methods can be considered: grafting of a compound on the fiber or surface modification by means of discharges.

The importance of plasma treatment has determined new methods to measure modifications entailed on films. An easy way of evaluating surface properties is by contact-angle measurements. A recently developed method for contact-angle measurements of a liquid drop on a thin monofilament has led to a better understanding of interactions between plasma and textiles [1].

This chapter surveys the surface characteristics of textiles treated by plasma for a range of fibers, with the exception of wool. Keratinous fibers have been studied and reported in many references [2,3].

Although the surface represents a small part of the fiber mass (1–2%), the level of its quality is decisive for many operating processes and for its utilization in a textile product. Fiber–polymer surface characteristics have an influence on the mechanical and chemical processing of fibers, yarns, and fabrics. For any fiber type, a plasma treatment changes one or more of the following qualities: water absorption and wetting [4–9] adhesion [4,6,10–13], dyeing and dye fastness [7,14], shade depth of dyed fabrics [15], antistaticity desizing [16], water and oil repellence [7,17,18] soil release [6,19], wear and chemical resistance [6], biocompatibility of the synthetic material with blood [20–26], and so forth. The use of discharges as a method for surface improvements represents both a source of theoretical information on polymer behavior and an uncommon and captivating engineering domain whose limits are not yet exactly known.

The interactions with fibers can be divided into chemical reactions, morphologic structure modifications, and polymer-forming reactions. In the improvement of fiber surface characteristics, a plasma can be used as one of the following types:

1. "Ambient" discharges such as corona
2. Reduced-pressure direct current, microwave, or radio frequency (RF) plasma
3. ion-beam discharges

Each has its advantages and disadvantages. Types 2 and 3 are more sophisticated in use but show a stronger and more stable modification of the polymer surface.

Research on applications of plasma discharges at atmospheric pressure revealed the possibility of using some special equipment to improve wettability, wickability, printability, and the surface contact angle of polymeric fabrics. Steady-state glow discharges at atmospheric pressure in helium, argon, air, carbon dioxide, and other gases [27,28] have been developed.

An ion-beam plasma has two plasmas: one used as an ion source and a second one for exposure samples to be treated. Despite the fact that some effects are similar, changes made by ion beam are more stable in time than modifications made by RF equipment and the same effect could be obtained at a shorter treatment time compared to RF plasma treatments.

A treatment [29] with a remote plasma on polyamide-66 (PA-66) and polyethylene terephthalate (PET) using nitrogen as the plasma gas has been reported. The sample is located in the afterglow region to avoid strong radiation, electric field, and collisions with charged particles. A very rapid, of the order of seconds, incorporation of nitrogen into polymers has been achieved, with the creation of some new functional groups (supposed as amides, imines, or nitrils).

In the case of RF discharges [30–33], the mechanisms of the surface phenomena are more complex because either the variety of ions and radicals impinging the dielectric surface is very large or their energy spectrum is very broad. Due to mobility differences, an electrically insulated surface exposed to the plasma develops a negative bias because of the greater mobility of electrons compared to positive ions. In this field, oriented positive-ion bombardment plays an important role in the plasma treatments of polymers. The following effects [31,34] determine the surface modifications of fibers: free-radical creation, functionalization, cross-linking, and degradation. A plasma creates these effects through energetic positive-ion bombardment of surfaces and generation of chemical active species. Ion energies play a main role in etching, but the degree of ionization and the production of activated neutral species have influence on surface modifications also.

The distribution of chemical species is a complex function of several parameters but depends mainly on the electron energy [31]. Usually, in oxygen plasmas, the negative atomic ions have no direct action on the polymer surface. The dissociation of O_2 is the most favored reaction which yields atomic oxygen; but the less probable atomic oxygen ionization can also take place, forming the O_2^+ molecular ion. In the case of a nitrogen plasma, the ionization energy of the N_2 molecule and N atom are higher than the plasma energy usually obtained in the polymer field, but an appearance of a metastable state (6.2 eV) is possible [31]. In an argon plasma, the active species are Ar^+ and argon in a metastable state (11.5 eV) [31]. Atomic fluorine and $CF_2-CF_2^+$ excited states have been detected [35] in CF_4 plasma.

Using a pure gas, one obtains specific plasma particles and behavior on a fiber. For a mixture of gases, the discharge becomes a very complex system. For different ratios of a two-gas mixture (CF_4/CH_4), there are concentrations where the etching is predominant and others where the polymerization process becomes important [36]. The ratio between gases of a mixture determines the stoichiometry of the active species in the discharge [37].

The difference in action of different plasmas consists also in the radiative content of the plasma—especially in ultraviolet (UV) radiation—which determines the appearance in the deeper layers of free radicals, reticulation, photolytical modifications, and so forth. Significant photolytical reactions in a 5–10-μm layer were observed on PET and polyethylene (PE) surfaces with cross-linking, change in crystallinity, and radical formation [38] as the main effects. A strong influence has been revealed concerning the interaction between polypropylene (PP) and PE with vacuum UV radiation [39].

In some treatment conditions, a cotton fiber surface shows chemiluminescence (CL) attributed to (1) activated carbonyl groups [40] in which the C atoms have a binding energy higher than in ground-state carbonyls or (2) reaction of free radicals with oxygen. Electron Spectroscopy for Chemical Analysis (ESCA) proved the appearance of aldehyde or keto groups. Those reagents that destroy free radicals are those that increase the CL of treated samples [40].

The plasma–fiber reactions take place by free-radical mechanisms and the fibers then undergo chain scission, radical transfer, oxidation, and recombination, all giving rise to a combination of degradation, ablation, oxidation, and cross-linking processes.

The surface functionalization induced by plasma is emphasized by the increase of the polar component of the surface energy. This polar function is mainly determined by oxygen even if the treatment is made in an inert gas. This specific feature has been explained by the existence of oxygen as an impurity in the plasma equipment or linked by adherence to the polymer or even the reaction of long-lived radicals on the sample with oxygen from the air. Water adsorbed on the inner walls of the reactor can also play the role of an oxygen source.

II. THE ROLE OF RADICALS

When fibers are exposed to plasma, their electron spin resonance (ESR) signals rise, showing the existence and reaction of radicals in plasma discharges. Their formation on the polymer surface is possible through hydrogen abstraction and through $C-C$ bond scission, after the bombardment with plasma particles [41]. Their appearance in a plasma system can be studied by the radical scavenger method [31,42]. Wakida et al. [43] carried out a study on radicals from miscellaneous fibrous polymers treated in different plasmas, achieving an interesting insight on the stability of free radicals. The ESR spectra for cotton samples irradiated in an oxygen plasma exhibited a less intense signal than those treated with a CF_4 plasma. In the case of the CF_4 plasma, the cotton signal is more intense than that of viscose rayon. The intensity of the radical signal of cotton treated by CF_4 plasma increases with the duration of the plasma treatment. A scale of radical intensity [43] has been published, confirmed in many parts by other research [44]. The signal intensity decreases in the order:

cotton > wool > silk > polyamide-6 (PA-6) ≅ PET

cotton > linen > mercerized cotton > polynosic fiber ≅ viscose rayon

CF_4 > CO > hydrogen > argon > CH_4 > nitrogen ≅ oxygen

From the experimental data, the polymer response to different plasma gas types depends on their chemical nature. Thus, the natural fibers produce more stable free radicals in comparison to synthetic and regenerated fibers. Also, it has been established [45] that PET is less affected by plasma than cellulosic products (mercerized cotton, viscose rayon, etc). The stability of free radicals is directed by the nature of the chemical bonds that are broken. In the case of cellulose, the appearance of the following free radicals [45] in the glucopyranose ring was suggested:

$$-\overset{\overset{\displaystyle H}{|}}{\underset{\underset{\displaystyle H}{|}}{C}}., \quad -\overset{\overset{\displaystyle H}{|}}{\underset{\underset{\displaystyle H}{|}}{C}}-O., \quad -\overset{\overset{\displaystyle H}{|}}{\underset{|}{C}}-O.$$

The type of the plasma gas plays a certain role on the radical formation: CF_4 and CO plasmas generate many more radicals in comparison to nitrogen and oxygen plasmas. For cellulosic fibers, the stability of free radicals is related to the fine structure of the fiber and the relative intensity increases with the duration of plasma treatment. From the relation among the intensity of free radicals, surface modification, and ESCA analyses [43], it was found that a CF_4 plasma treatment determines the surface modification for any cellulose fiber in a similar way. For cellulosic fibers, there are no links between the radical intensity and the magnitude of surface modification. In the case of regenerated fibers, the formed radicals recombine very fast due to the loose substrate structure. Research done on the I_2 plasma has revealed that iodine atoms are bound chemically to the surface of PE [46].

A test conducted by Yasuda [47] consisted of a glass rod coated with PE films exposed to nitrogen RF plasma. The ESR spectra showed the existence of free radicals in the PE layer and also in the glass, which was not reached by the plasma.

It seems that for RF plasmas, an increase of power implies a diminution in radical content both for oxygen and argon plasma treatments [31]. The increase in RF power increases the kinetic energy and the number of particles that reach the polymer surface. In the case of scission and cross-linking processes, the first one becomes more probable. The radical content dependence upon treatment time could have the same explanation. The lower energy and density of the active particles in plasma promote a depth effect. For treatments in an oxygen plasma, the increase in the gas flow rate implies an increase in radical content. The increase of the argon flow rate allows a lowering of the radical content [31].

The first steps in chemical reactions promoted by UV radiation and atomic oxygen on PET are governed by the following radicals [48]:

$-PhCOOCH_2CH_2-$

$$
\begin{aligned}
&\rightarrow (a_1)\; -PhCOO\cdot + \cdot CH_2CH_2- \quad \rightarrow (b_1)\; -PhCOOH + \cdot CH_2CH_2- \\
&\qquad\qquad\qquad\qquad\qquad\qquad\quad \rightarrow (b_2)\; CO_2 + -Ph\cdot + \cdot CH_2CH_2- \\
&\rightarrow (a_2)\; Ph\cdot + \cdot COOCH_2CH_2- \quad \rightarrow (b_3)\; CO_2 + -Ph\cdot + \cdot CH_2CH_2- \\
&\rightarrow (a_3)\; PhCO\cdot + \cdot OCH_2CH_2- \quad \rightarrow (b_4)\; CO + -Ph\cdot + \cdot OCH_2CH_2-
\end{aligned}
\tag{1}
$$

where Ph represents a phenyl group. Then, the atomic and molecular oxygen species attack the primary and secondary polymer radicals and induce the propagation of the degradation until the final products of destruction, CO, CO_2, H_2, H·, and ·OH, are obtained [49].

III. TRENDS IN THE CROSS-LINKING OF A POLYMER SURFACE

As an effect of the action of free radicals, there is the possibility of cross-linking. The presence of a cross-linked network can be investigated by polymer dissolving tests. Other effects are a decrease of the thermo-oxidative decomposition rate and an increase of the molecular weight. Both reactive and inert plasma gases [31] induce, at the same time, a cross-linking and a chain scission process, but the treatment conditions seem to determine which one prevails. The presence of oxygen traces in a plasma inhibits the cross-linking process by the appearance of new functional groups on the PET chains as well: hydroxyl, hydroperoxyl, carbonyl, carboxyl, and so forth.

In the case of PET cross-linking [50,51] in the first stage, hydrogen is abstracted from the benzene ring, after which a recombination of two aryl radicals occurs:

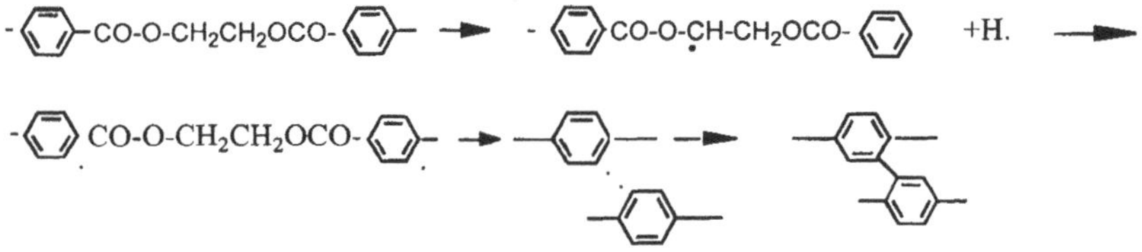

IV. THE SURFACE DYNAMICS OF FIBROUS POLYMERS

The treated fibers show the tendency to adjust their surface in agreement with the energy value of the outside medium [52]. Once in contact with air, considered as a medium with low surface energy, the polar groups from the surface embed into the matrix. In contact with water, as a medium with high surface energy, the polar groups remain exposed to the interface, which blocks their arrangements and promotes them in a thermally equilibrated surface [52]. The adjustment in water of the polymer surface treated by plasma is at least two orders of magnitude longer than the time needed to measure contact angle in the interface with air [52]. These modifications do not return to the value before water treatment because in spite of macromolecular motions, the subsurface polymer structure is not able to embed all functional groups away from the surface. Experiments with ^{18}O and ^{16}O have demonstrated [52], at least for the PP treatment by oxygen plasma, that the insertion of oxygen due to reactions of active sites with atmospheric oxygen is negligible. This has been confirmed by other research [53]. Plasma-treated PP samples consist of two different layers. The bulk is untreated, but the topmost layer, which has a thickness of less than 5 nm, consists of a random copolymer of oxidized and unchanged units. This layer rearranges itself by thermally activated macromolecular motions in a way that depends on the medium interfaced, toward a minimum in interfacial energy. (i.e., surface tension in physicochemical terms).

During research in the dynamics of free surface energy in time, a geometric mean approximation has been made using two liquids with known surface free energy, whereby all interactions are treated without differentiating among dipole–dipole and acid–base interactions and hydrogen-bonding. In Ref. 48, the free surface energies for the used polymers are decomposed into terms arising from dispersive interactions, more exactly considered Lifshitz–van der Waals interactions [48]. After the plasma treatment, the polar component remains greater than the value for before treatment.

The adjustment capacity of polymers such as PA-6, PA-66, PET, and PP films, or fibers to the level of environmental energy can be undesirable for the stability of wettability, adhesion, and mechanical properties during use. PP films treated by ammonia plasma showed a decrease of wettability and adhesion [4,31] in time. In order to improve the behavior of the surface layers, it is necessary to cross-link them.

The aging process could be avoided by strengthening the cross-linked layer using a He or Ar plasma. CASING, the acronym for "cross-linking by means of activated species of inert gases," which defines a process of inserting new chemical bonds including unsaturated ones, increases the mechanical resistance of the surface by formation of a very cohesive and denser layer [4]. Concerning this point of view, a noble gas plasma pretreatment could be used in order to minimize the surface rearrangement of plasma treated PP on contact with ambient air. In this way, the role of the two plasma treatments to immobilize the top surface layers has been stated [4] by measurements of elongation at break. Then, the adhesion properties of the surface are prepared. These conditions have been achieved by a treatment in a He plasma, followed by a treatment in an ammonia plasma. Tests were carried out on the samples treated by ammonia plasma with or without a pretreatment of He plasma. Elongation at break measurements showed that without pretreatment, the adhesion degraded, whereas a He pretreatment caused a stable and good adhesion. An optimal treatment time of 1 s has been found, and even in the case of the longer treatment times, good results were obtained.

Tatoulian et al. [55] demonstrated the appearance of polar basic groups at the PP surface. The basic character was no longer observed with aging and the surface has an amphoteric character instead. In the presence of the He plasma pretreatment, the surface is kept almost unchanged. The surface layer of NH_3 plasma-treated PP films with a polar group content is less rapidly rearranged when doing a He plasma pretreatment.

Auger electron spectroscopy (AES) was used to identify chemical bonds at an aluminum–polypropylene interface in a study of the behavior of thin aluminum film formed by thermal evaporation in a modified plasma chamber. For He-pretreated PP samples, AES indicated a sharp interface showing no aluminum interpenetrating with the cross-linked layer of the surface. In the absence of a He pretreatment, the samples treated by NH_3 plasma revealed a broader interface. The He plasma pretreatment seems to cause an outstanding stabilization of the PP surface through a lower mobility of polymer chain segments, which is correlated with cross-linking density.

On the other hand—and opposed to polymer cross-linking—chain fragments can be formed either at the surface or deeper in the polymer volume, caused by vacuum UV radiation. Fragments can migrate to the surface and, on their way, affect interface properties. The higher power of the helium plasma discharges causes a great fragmentation of the polymer surface, which causes a higher mobility of the polar groups. Thus, surface cross-linking occurs simultaneously with chain scission. Polarity increases and the amount of polymer fragments rises with plasma power [4]. The formation of fragments at the polymer surface can lead to a weak boundary layer, which causes the surface to lose its properties with aging time.

When polyurethane (PU) used in biopolymer applications is immersed in orienting fluids, a significant increase takes place in the polar component of the free surface energy. This response is a function of the immersion medium's acid–base interaction potential. The restructuring from the as-cast state is due to a preferential concentration of low-energy segments on the surface. To avoid a change in surface properties for biological applications, a successful attempt led to cross-linking the surface by means of microwave plasma discharges in argon medium. The PU surface modified responded much more weakly to changes in the polarity of contact media [54].

V. THE ETCHING PROCESS

The etching process in plasma discharges implies the collision between the fiber surface and particles from the plasma medium, thus removing the top layer of the fiber. Etching can be done on the whole surface for cleaning or for dimensional modification, or on location in a pattern, limited part of the surface being etched by the discharge.

The surface can undergo a large series of modifications: creation of simple vacancies, and microcavities, affectations of the stoichiometry reflected in the level of the electric charge, and so forth [8]. The shifting of stoichiometry generates fiber changes and the possibility of the emission of a single atom; this process is called sputtering. In this area of interactions, both ions and atoms can participate, but the influence of ions is greater because they can be accelerated in the electric field of the plasma.

The selective etching occurs depending on local densities as a result of uneven packaging of the polymer chains. After plasma etching, scanning electron microscopy (SEM) images show the crystalline zones of higher density as light spots in a dark field which consists of the amorphous parts of lower density. Because density differences can be extremely small, the surface degradation under a particle flux has to be monitored carefully. The particle energy must have an optimum value to selectively etch the domain of lower density while avoiding the degradation of the crystalline areas. This etching process is a visualization method for polymer structure by means of SEM and atomic force microscopy (AFM). Chemical etching is not always selective enough to obtain high resolution. As an example, the etching with phenol/tetrachloroethane is not selective. The use of chromic acid or of bases is selective but not to the same degree as plasma etching, provided the operating conditions are well chosen. When the energy is too high, particles do not act selectively and the differences between the crystalline and amorphous regions cannot be observed. An increase of the etching time affects only the degree of etching; the selectivity remains unchanged. The shape of the substrate has an influence on the effect: film, fiber, or powder. To obtain SEM replicas, there is a minimum etching time needed to visualize the characteristic globular aspect of a polymer surface: 5 min in oxygen and around 15 min in nitrogen and argon [32].

Concerning etching sensitivity, polymers which contain oxygen groups show a higher aptitude to etching, unlike the hydrocarbon polymers. At constant plasma parameters, the synthetic fiber most sensitive to the discharge etching seems to be PET, and the least sensitive is PP, in terms of weight loss rate. As an example [56], the rate of weight loss decreases in the following order: polyoxymethylene > poly(vinyl alcohol) > PET > PA-6 > PP.

VI. THE SPECIFIC BEHAVIOR OF TEXTILES

A study carried out [57] on the modifications of plasma-treated PA-66, PA-6, and PET fibres reported some aspects determined from morphology changes by means of AFM images.

One considers that synthetic fibers are lacking the advantages which the existence of microporosity can induce on superficial properties. A representative synthetic fiber such as PA-66 was examined by visualizing the morphological surface details with the AFM technique and by sampling longitudinal sections with a laser at a depth of 2–3, 5–7, and 10–17 nm. A comparison of the AFM images of a nontreated PA-66 surface (Fig. 1) with a PA-66 surface treated with an air or argon plasma (Figs. 2 and 3, respectively) for the same depth level resulted in a significant restructuring which evidences the differentiated action of an argon plasma as compared to an air plasma.

The images obtained for fibers treated in air plasma show micropores uniformly arranged, with diameters of about 40 nm and maximum lengths up to 500 nm. For the fibers treated in argon plasma, with samples taken under the same conditions, there is a tendency toward pores with a larger diameter (130 nm) and a reduced frequency of micropores, which are surrounded by large amounts small-dimension porosities (up to 20 nm). On comparing the obtained images, the action of argon plasma as an etching medium is more efficient than an air plasma in what the porous amplification is concerned. In this process, there is an implication of the partial crystalline zones next to the amorphous ones.

In agreement with the porosity, the phenomenon of water and dye diffusion can be dealt with by estimating the size of the dispersed dyes with which the fibers interact. For this class of dyes, the study started from their general chemical

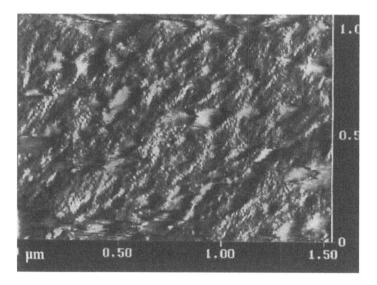

FIG. 1   AFM image of PA-66 untreated.

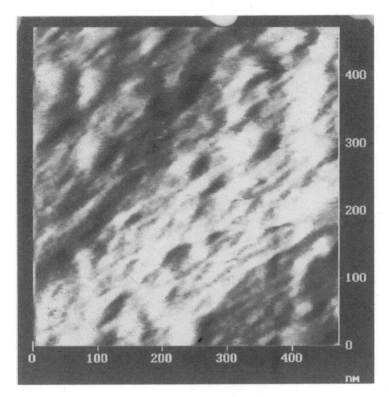

FIG. 2 AFM image of PA-66 treated with air plasma.

structure [58]. The most unfavorable diffusion case was considered, where the dye has in its molecular structure the most voluminous substituents.

With the help of Alchemy II software, it was determined that the size of the investigated dispersed dye molecule is always smaller than 2 nm.

The comparison of the pore size from the AFM images with the dye dimension brings about the possibility of dye diffusion—even in the solid state—through the created pores into the fiber. According to data reported in Ref. 59 for a pore size of about 1 nm, the diffusion is of configurational type; for a pore size in the range 2–10 nm, it is of the Knudsen type, and for even larger pores it is of regular type.

From the estimations of the maximum values of pores which offer the available space for water and dye transfer from the outside toward the inside of the matrix, one establishes that the global transfer coefficient of the various penetrants is the sum of more types of diffusion. Microporosity as a result of plasma

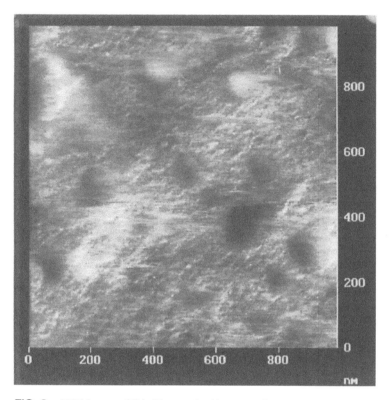

FIG. 3 AFM image of PA-66 treated with argon plasma.

action allows the appearance of a dual diffusion mechanism: via free volume and via pores. The diffusion of water and dye are improved, also the water sorption.

Investigating the working depth of a plasma on porous materials as well as textiles, Yasuda [60] stated that plasma irradiation does not penetrate into the inner part of a small-diameter tube. However, less reactive species, controlled by diffusion, could penetrate deep into the tube, whereas highly reactive species only react at the entrance. Treatments with CF_4 and C_2F_4 plasmas penetrate to a considerable thickness into the fibrous material, depending on the porosity of the system.

Silk fibroin membranes have been fluorinated by CF_4 plasma (gas flow rate 30 mL/min; 2 min; 30–400 W; 13.56 MHz), followed by immersing in impure water. The hydrophobicity and the water contact angle increased and the swelling capacity decreased with the increase of the fluorine content. The formation of primary amino groups by the decomposition of peptide bonds was suggested because of an increase of the dyeing capacity with acid dyes [17].

Cotton treated in an oxygen plasma revealed reduced strength, but the rate of weight loss in a subsequent enzyme (cellulase) treatment decreased compared to untreated cotton. Fabrics treated by plasma showed a lower equilibrium dye uptake due to the cross-linking of the cotton surface [61]. The treatment also reduces the susceptibility of cotton to cellulase and dye molecules. The cellulase treatment of cotton fabrics increases the dyeing rate, but decreases the equilibrium dye uptake by digestion of the amorphous regions of the fiber. One can note that plasma modifies the surface of a fiber, and the enzyme affects the inner part of the polymer primarily.

The PA-6 fibers were treated with O_2, CF_4, or He/Ar plasma and then dyed with acid and basic dyes [62]. There was a decrease of acid dye diffusion and an increase of the basic dye diffusion. As a result of oxygen plasma, much oxygen is incorporated onto the fiber surface in the form of $-OH$ and $-COOH$ groups, which increase the negative charge of the fiber surface [46]. Compared to non-treated fibers, plasma-treated fibers show a lower acid and higher basic dye sorption for the same reason.

Silk, PA-6 and chinon (promix fiber) fabrics dyed with Cl Acid Black 155 were sputter etched in the presence of argon etched (20 W; 0.1 torr; 10–300 s) and then treated in an argon plasma under the same conditions. They show an increase in the depth of shade. The small microcraters which formed during treatment considerably reduced light reflection [15]. The formation of the microcraters smaller than the wavelength of visible light plays an important role on the depth of shade of black-dyed PA-6 fabrics.

The dyeing kinetics of PP fibers dyed with Resolin Red FB improved for textiles pretreated with an O_2 plasma. Klein et al. [63] considered the improvement of dyeing behavior to be determined by the surface enlargement through etching, by the appearance of new functional groups able to establish additional links between PP and dye molecule, as well as hydrophobic interactions and van der Waals forces, and by improvement of wettability.

If the plasma acts preferentially on noncrystalline domains, one can presume a diminishing dyeability. If the discharge etching opens up new available regions for the dye sorption, this would entail an enhanced dyeability. After a treatment, the fiber's surface becomes activated and the dye sorption on the inside of the interface usually improves. One can consider that the incorporation of fluorine by means of CF_4 on PET and PA-6 films occurs independently of the crystallinity of the samples. Chemical reactions determined by plasma particles take place with all molecules at the surface [64]. Reports on PET and PA-66 treated by RF air plasma indicate fibers with significantly reduced dyeability [14]. The crystalline and quasicrystalline regions are least disturbed. On the other hand, plasma treatments in an oxygen medium improved the surface dyeability of PET fabrics [7]. A first aspect is linked to diffusion of the dye toward the surface, which could be improved through electrostatic charge attraction or by polar groups able to establish hydrogen bonds, or diminished by repulsion of like charges. A second aspect refers to the amount of dye retained by fibers, which is always lower compared to nontreated fibers, because etching especially damages the amorphous areas of fibers where normally dyes penetrate.

REFERENCES

1. N. Dumitrascu, C. Agheorghiesei, and G. Popa, Entropie *213*(2):9 (1998).
2. W. Rakowski, in Plasma Treatment of Wool, Biella Wool Textile Award, Italy (1992), pp. 1–57.
3. K. M. Byrne, W. Rakowski, A. Ryder, and S. B. Havis, presented by K. M. Byrne, Proc. IX International Wool Textile Res. Cont., Biella, Italy, 1995, pp. 234–243.
4. M. Tatoulian, F. Arefi-Khonsari, I. Mabille-Rouger, J. Amouroux, M. Gheorghiu, and D. Bouchier. J. Adhes. Sci. Technol. *9*(7):923 (1995).
5. T. Wakida, H. Kawamura, J. C. Song, T. Goto, and T. Takagishi. Sen-i Gakkaishi *43*:384 (1987).
6. C. Tomasino, J. J. Cuomo, C. B. Smith, and G. Oehrlein. J. Coated Fabrics *25*:115 (1995).
7. A. M. Sarmadi and Y. A. Kwon. Textile Chem. Color. *25*(12):33 (1993).
8. C. D. Radu, Ph. D. thesis, Rotaprint, Iasi, Romania, 1995.
9. C. D. Bain and G. M. Whitesides. J. Am. Chem. Soc. *110*:5897 (1988).
10. M. Shaker, I. Kamel, F. Ko, and J. W. Song. J. Compos. Tech. Res. *18*:249 (1996).
11. J. Jang and H. Kim. Polym. Compos. *18*:125 (1997).
12. M. Strobel, C. S. Lyons, and K. L. Mittal (eds.), *Plasma Surface Modification of Polymers: Relevance to Adhesion*, VSP, Zeist, The Netherlands, 1994, pp. 17–34.
13. F. Arefi, V. Andre, P. Montazer-Rahmati, and J. Amoroux. Pure Appl. Chem. *64*: 715 (1992).
14. T. Okuno, T. Yasuda, and H. Yasuda. Textile Res. J. *62*(8):474 (1992).
15. J. Ryu, J. Dai, K. Koo, and T. Wakida. J. Society of Dyers and Colorists *108*:278 (1992).
16. F. Valu and C. D. Radu, presented by C. D. Radu, Symposium Ses. Jub. Com. St., Iasi, Romania, 1988.
17. A. Nishikawa, H. Makara, and N. Shimasaki. Sen-i Gakkaishi, *50*(7):274 (1994).
18. Y. Iriyanma, T. Yasuda, D. L. Cho, and H. Yasuda. J. Appl. Polym. Sci. *39*:249 (1990).
19. C. J. Jahagirdar and V. Atakrishnan. J. Appl. Polym. Sci. *41*:117 (1990).
20. N. Dumitrascu, V. Stanciu, and G. Popa, Balkan Physics Letters *48*:1909 (1997).
21. B. D. Ratner. J. Biomater. Sci. Polym. Ed. *4*(1):3 (1992).
22. N. Dumitrascu, S. Surdu, and G. Popa, presented by N. Dumitrascu ESCAMPIG, Poprad, Slovakia, 1996, p. 20 E:395.
23. S. Surdu, N. Dumitrascu, C. Avram, and G. Popa, presented by N. Dumitrascu, 11th Balkan Biochemical Biophysical Days, Thessaloniki, Greece, 1997, p. 65.
24. M. Gheorghiu, G. Popa, and O. C. Mungiu. J. Bioact. Compat. Polym. *6*(2):164 (1991).
25. E. Ruckenstein and V. Sathyamurthy. J. Colloid Interf. Sci. *101*:436 (1984).

26. M. Sotton and G. Nemoz. J. Coated Fabrics *24*:139 (1994).

27. S. Tokino, T. Wakida, H. Uchiyama, and M. Lee. J. Society of Dyers and Colorists *109*:334 (1993).

28. P. P. Tsai, L. C. Wadsworth, and J. R. Roth. Textile Res. J. *67*(5):359 (1997).

29. Foerch and H. Hunter. J. Polym. Sci. Part A: Polym. Chem. *30*:279 (1992).

30. C. K. Birdsall and V. Vahedi, presented by R. Schrittwieser, Fourth Symp. on Double Layers and Other Nonlinear Potential Structures in Plasmas, Innsbruck, 1992.

31. C. Popa and M. Gheorghiu. Roman. Rep. Phys. *46*:307 (1994).

32. J. P. Boeuf, P. Bellenguer, L. C. Pitchford, and I. Peres, Courses CIP 91, Antibes, 1991.

33. P. A. Chatterton, J. A. Rees, W. L. Wu, and K. Al. Assadi. Vacuum *42*:489 (1991).

34. A. Schonhorn and R. H. Hansen. J. Appl. Polym. Sci. *11*:1461 (1967).

35. J. Bretagne, F. Epaillard, and A. Ricard. J. Polym. Sci. Part A: Polym. Chem. *30*: 323 (1992).

36. X. Wei, C. Xiaodong, and W. Jianqi. J. Polym. Sci., Part A: Polym. Chem. *33*:807 (1995).

37. J. W. Coburn and A. F. Winters. J. Vac. Sci. Technol. *16*:391 (1979).

38. J. Frydrich, G. Kuhn, and J. Gahde. Acta Polym. Chem. Ed. *17*:957 (1979).

39. A. Hollander, J. E. Klemberg-Sapieha, and M. R. Wertheimer. J. Polym. Sci. Part A: Polym. Chem. *33*:2013 (1995).

40. R. R. Benerito, T. L. Ward, D. M. Soignet, and O. Hinojosa. Textile Res. J. *57*:224 (1981).

41. N. Morosoff, B. Crist, M. Bumgarner, T. Hsu, and H. Yasuda. J. Macromol. Sci. *A10*(3):451 (1976).

42. H.-U. Poll, R. Kleemann, and J. Meichsner. Acta Polim. *32*:139 (1981).

43. T. Wakida, K. Takeda, I. Tanaka, and T. Takagishi. Textile Res. J. *59*(1):49 (1989).

44. J. R. Chen. J. Appl. Polym. Sci. *42*:2035 (1991).

45. C. I. Simionescu, F. Denes, M. M. Macoveanu, and I. Negulescu. Makromol. Chem. *8*(Suppl.):17 (1984).

46. M. Fujishima, D. Kawabata, C. Funakashi, Y. Yoshida, T. Yamashita, K. Kashiwagi, and K. Higashikata. Polym. J. *27*:575 (1995).

47. H. Yasuda. J. Macromol. Sci. Chem. *A10*(3):383 (1976).

48. M. Gheorghiu, F. Arefi, J. Amouroux, G. Placinta, G. Popa, and M. Tatoulian. Plasma Sources Sci. Technol. *6*:8 (1997).

49. F. Clouet and M. K. Shi. J. Appl. Polym. Sci. *46*:1955 (1992).

50. M. Macoveanu, Ph.D. thesis, Inst. Chim. Macro., P. Poni, Iasi, Romania, 1981.

51. F. Gugumus, *Plastic Additives Handbook*, Hanser, Munich, 1990, Chap. III.

52. E. Occhiello, M. Morra, G. Morini, F. Garbassi, and P. Humphrey. J. Appl. Polym. Sci. *42*:551 (1991).

53. Th. Gross, A. Lippitz, W. E. S. Unger, J. F. Friedrich, and Ch. Woll. Polymer *35*: 5590 (1994).

54. K. Tingey, K. Sibrell, K. Dobaj, K. Caldwell, M. Fafard, and H. P. Schreiber. J. Adhesion, *60*(1–4):27 (1997).

55. M. Tatoulian, N. Shahidzadeh, F. Arefi, and J. Amouroux. Int. J. Adhesion Adhesives *15*:1212 (1995).

56. N. Inagaki, in *Plasma Surface Modification and Plasma Polymerisation*, Technomic, Lancaster, PA, 1996, p. 23.
57. C. D. Radu, P. Kiekens, J. Verschuren, and N. Asandei, presented by C. D. Radu, Proceedings of ArchTex'97 Textile-Designing & Care. Institute of Textile Architecture, Lodz, Poland, 1997.
58. T. Forst, in *Chimia Colorantilor*, Rotaprint, Iasi, Romania, 1987, p. 34.
59. P. B. Venuto. Micropor. Mater. *2*:297 (1994).
60. T. Yasuda, T. Okuno, M. Miyama, and H. Yasuda. J. Polym. Sci. Part A: Polym. Chem. *32*:1829 (1994).
61. N. S. Yoon, Y. J. Lim, M. Tahara, and T. Takagishi, Textile Res. J. *66*:329 (1996).
62. T. Wakida, M. Lee, S. Niu, S. Kobayashi, and S. Ogasawara. Sen-i Gakkaishi *50*: 421 (1994).
63. C. Klein, H. Thomas, and H. Hocker, presented by C. Klein, Aachener Textiltagung, Okonomischer Gewinn aus okologischer Optimierierung, Aachen, 1997.
64. T. Yasuda, T. Okuno, K. Yoshida, and H. Yasuda. J. Polym. Sci. Phys. *26*:1781 (1988).

9

Measuring Interface Strength in Composite Materials

PETER SCHWARTZ   Department of Textiles and Apparel, Cornell University, Ithaca, New York

I. INTRODUCTION

The key to the performance and properties of composite materials is the ability of the matrix to transfer load, through shear, to the reinforcing fibers. A knowledge of the mechanics and mechanisms of fiber–matrix adhesion leads to a better understanding of its role in composite strength and to better design and utilization of composites for appropriate applications.

At first, it was believed that the load transfer from matrix to fibers occurred precisely at the interface between the two. Presently, as a result of many carefully controlled experiments and computer analyses [1–5], there exists very strong evidence for the presence of a three-dimensional region, or interphase. This region extends a short distance, on the order of hundreds of nanometers, into both the bulk matrix and the fiber, as schematically illustrated in Fig. 1.

Because of its importance, measurement of the interface strength is important for the prediction of the overall mechanical properties of the composite material. There are two general approaches to performing this measurement. In the first case, the strength is measured using a single fiber embedded in a matrix; without other surrounding fibers. In these tests, the fiber volume fraction, V_f, is essentially zero. The second approach, an attempt to replicate more closely a fiber in a composite, uses a fiber surrounded by other fibers, or an actual unidirectional composite specimen. For single-fiber techniques and many multiple-fiber techniques, however, assuming a uniform stress profile along the fiber's length, the calculation of interfacial shear strength (IFSS), τ, or, more precisely, the average interfacial shear strength, is generally found using a simple force balance first reported by Kelly and Tyson [6]:

$$\tau = \frac{P}{\pi dl} \tag{1}$$

where P is the maximum force required to produce fiber movement, d is the fiber diameter, and l is the embedded fiber length.

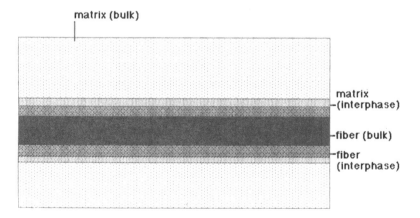

matrix (bulk)

matrix (interphase)

fiber (bulk)

fiber (interphase)

FIG. 1   Interphase region in a fiber-reinforced composite.

II. SINGLE-FIBER TECHNIQUES

A. Fiber Pullout

The oldest [7] single-fiber technique used to determine IFSS is the fiber pullout test [8] (Fig. 2), wherein the fiber is pulled from a bulk coupon of matrix; an ideal load trace is illustrated in Fig. 3. At the beginning of the test, the load on the fiber builds until it reaches a peak and the interface between the fiber and the matrix fails. The load immediately drops, but not necessarily to zero if a sliding friction force is encountered as the now debonded fiber is pulled out from the matrix.

One major consideration in using this procedure is the geometry of the matrix portion. First, the thickness, l, must be such that the pullout force does not exceed the fiber's breaking strength, otherwise the matrix will act as a bottom jaw in a simple tensile experiment. Second, the sample should be prepared so that it is easily held, and both small coupons and thin disks have been used. Li and Netravali [9] have proposed a coupon geometry where projections from the main body are gripped, rather than the coupon body itself, in an attempt to eliminate effects from the transverse jaw force. Banbaji [10], using a polymer–cement system,

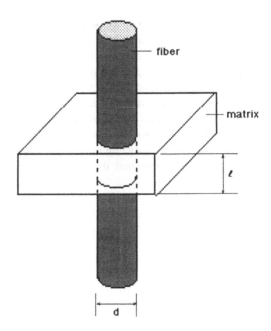

FIG. 2 Geometry of the fiber pullout test.

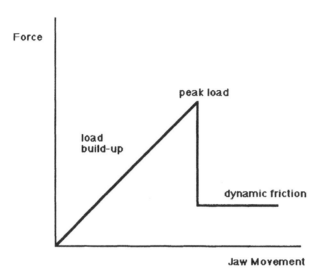

FIG. 3   Idealized force trace in a fiber pullout test.

has shown (Table 1) that this force can make a great impact on the measured pullout force, whereas Gray [11] and Piggot [12] noted an effect of embedded length on the interface strength. From Eq. (1), for the same epoxy–fiber system and for fibers having the same diameter, one should expect the pullout force to be proportional to the embedded length, $P \propto l$. Piggot's results [12], using a glass–low-density polyethylene (LDPE) system, suggested that at some embedded length, the pullout force is depressed from what one would predict using Eq. (1), with values of τ calculated from tests at shorter embedded lengths.

Because the embedded length is a critical factor in analyzing these test results, the meniscus formed at the point of entry of the fiber into the matrix by the

TABLE 1   Effect of Pressure on Maximum Pullout Force in a Polymer–Cement Composite

Jaw pressure (MPa)	Embedded length (mm)	Pullout force (N)
17.6	4.2	41.2
35.2	4.3	62.0

Source: Ref. 8.

epoxy's wetting the fiber, illustrated in Fig. 4, was of much concern. More refined techniques for sample preparation [9] have, to a large extent, reduced this problem; however, this concern was one of the factors leading to the development of another pullout technique—the microbond or microdroplet method.

B. Microbond Test

First reported by Miller et al. [13], the microbond test was proposed to solve two major problems encountered using pullout tests: meniscus effects and the difficulty of casting samples with sufficiently small embedded lengths. Using the microbond method, a droplet of matrix material is placed on the fiber and the fiber is pulled through a small opening (e.g., a hole in a plate, an opening between two parallel plates, etc.) causing the droplet to debond and slide along the fiber (Fig. 5). The force trace is identical to that obtained using the pullout test, and using the finite-element method (FEM) of analysis, it has been shown [14] that the two tests, performed correctly, are mechanically equivalent. In addition, it has been shown that the geometry of the droplet—spherical or ellipsoidal—is not a factor [15].

Because the microbond test is essentially a fiber pullout test, many of the problems that exist in fiber pullout techniques are also present when using the microbond technique. Care should be taken to produce droplets with minimal meniscus effects, grossly irregular geometries, and inappropriate embedment lengths. One difference between this test and the standard fiber pullout test is

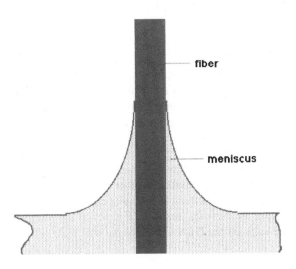

FIG. 4   Meniscus effect in a fiber pullout test.

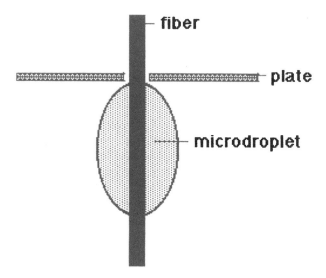

FIG. 5 Schematic of the microbond test.

that the droplet size (length) cannot be precisely controlled; rather, a range of sizes are usually present. Because the embedded length has been shown to influence the calculated interface strength [12], care is usually taken to use microdroplets within a narrow range of lengths. One advantage of the test is that if one is careful and uses a long enough fiber, the droplets may be spaced to permit multiple tests using the same fiber.

C. Single-Fiber Fragmentation

The first reported use of the single-fiber fragmentation technique was by Kelly and Tyson [6]. For this procedure, a single fiber is embedded in a matrix coupon, often in the shape of a dog bone. This "single-fiber composite" is then subjected to a uniform end stress, and the fiber, because of stress transfer in the matrix, begins to carry load (Fig. 6a). As the load is increased the fiber will fracture, and various flaw points. If the fiber fragment is long enough, the load on the fragment can increase if the applied load is increased, and it may fragment at several other locations (Fig. 6b). Below this length, the fiber fragment can carry no additional load. The test continues until saturation occurs i.e., each fiber fragment has length less than or equal to a critical length l_c (Fig. 6c). If the strength of the fiber at length l_c is known, then the IFSS can be found by

$$\tau = \frac{P_{l_c}}{\pi d l_c} \tag{2}$$

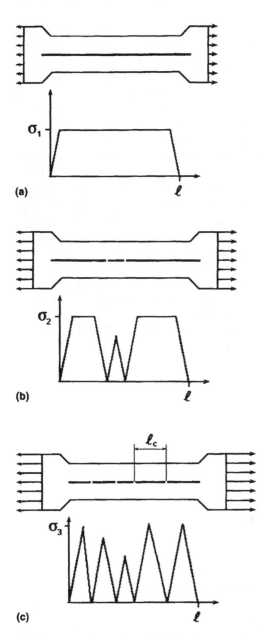

FIG. 6 Stress in the fiber during the fragmentation test: (a) before a fiber break; (b) after two breaks; and (c) at saturation.

Several issues must be addressed to apply the results of this technique to determine the strength of the interface. Because the procedure relies on the matrix remaining intact throughout the test, the failure strain of the matrix must necessarily be higher than that of the fiber; Kelly and Tyson [6] recommend that the matrix have at least three times the extension of the fiber.

A second issue is that the test relies on knowledge of the fiber's strength *at the critical length* l_c, typically on the order of 8–10 fiber diameters [16]. Given that typical fiber diameters range from 6 to 30 μm, l_c is much smaller than any length used to determine single-fiber tensile strength [17]. A common technique of determining this is the application of the Weakest Link Theorem due to Peirce [18]. The results of this theorem indicate that the relationship log(fiber strength) versus log(gauge length) should be linear, as illustrated in Fig. 7; the strength at shorter lengths (and, by symmetry, longer lengths) may be found from a knowledge of the strength at "reasonable" gauge lengths. (See Ref. 17 for an example single-fiber testing techniques and the application of the Weakest Length Theorem to *para*-aramid fibers.)

Another issue is that the test results are reliable only for fibers that show clean, brittle fracture (e.g., graphite and glass) rather than splitting (e.g., *para*-aramids), because the fiber fracture point is well defined, and fragment lengths can be measured. Generally, these tests are performed using transparent matrices, but,

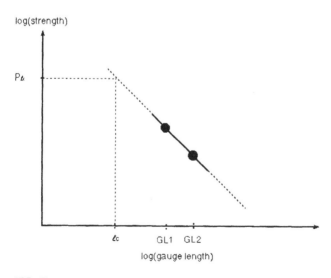

FIG. 7 Determination of breaking load at the critical length using the Weakest Link Theorem.

using acoustic emission [19], fiber fragmentation tests using opaque matrices may be performed.

Finally, it should be noted that at saturation, because weak flaws are randomly distributed along the fiber's length, the fragment lengths too are necessarily different. The question then becomes one of choosing an appropriate value for l_c. It has been argued [11,20] that because the flaws are randomly distributed, the fragment lengths at saturation should be uniformly distributed over the interval $[l_c/2, l_c]$, and so the appropriate value is

$$l_c = 0.75\bar{l} \tag{3}$$

where $\bar{l}$ is the mean value of the measured fragment lengths.

III. MULTIPLE-FIBER TECHNIQUES

A. Microindentation

The microindentation technique [21] is the "inverse" of the fiber pullout techniques; in this test, the fiber is pushed out of the matrix. Generally, the procedure is run using a highly polished, high-V_f-composite cross section. Individual fibers may be selected that are closely surrounded by neighboring fibers either in hexagonal or square array or in resin-rich areas of the composite. A microindenter is carefully placed in contact with the fiber, and a force, P, is applied until the fiber is pushed through the matrix (Fig. 8). The IFSS is calculated [6] using

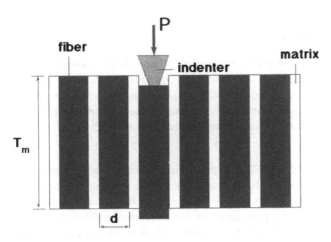

FIG. 8 Schematic of the indentation test.

$$\tau = \frac{\sigma}{2} \left(\frac{G_m}{E_f}\right)^2 \left(\frac{d}{T_m}\right)^2 \tag{4}$$

where σ is the load P divided by the contact area of the indenter, G_m is the shear modulus of the matrix, E_f is the elastic modulus (in compression) of the fiber, d is the fiber diameter, and T_m is the specimen's thickness.

Kalinka et al. [21] mention that the preparation of sufficiently thin specimens is difficult, and so microindentation is used primarily with composites having low interface strength and relatively thick fibers (e.g., ceramic–ceramic composites). Also, the fibers must be relatively brittle and not subject to splitting before being pushed through, as is the case when this technique is used with *para*-aramids or high-strength polyethylene (PE).

B. Single-Fiber Pullout from a Microcomposite

In an attempt to measure IFSS using fiber pullout techniques at a typical composite V_f, Qiu and Schwartz [22] described a technique in which a single fiber is pulled from a seven-fiber, matrix-impregnated, hexagonal array, illustrated schematically in Fig. 9. Termed SFPOM, the technique is designed to approximate the conditions a fiber would see in a real composite. Moreover, like single-fiber pullout tests, the average IFSS is be calculated using Eq. (1). For the same embedded length, Qiu and Schwartz found that the measured IFSS was lower using SFPOM than the microbond test (Table 2). They also found that the IFSS decreased with increasing V_f, which they attributed to shear-stress concentrations at the surface of the fiber and matrix failure resulting from softening of the matrix in the interstices between fibers.

C. Composite Laminates

Using composite laminates to determine interface strength has the advantage over other methods in that it uses real composites. Its chief disadvantages are that the results are more qualitative than quantitative and the samples more expensive to produce. Tension and compression tests are commonly used. The modes of failure in these tests provide a qualitative measure of interface strength.

1. Tensile Tests

Tensile tests (e.g., ASTM D3039-76) are used primarily to find the ultimate properties of composites as well as to determine the elastic constants of the fibers and the matrix. These tests also provide a clue as to the quality of adhesion between the fibers and matrix. Poor adhesion can be inferred when the specimen fails largely through with fiber pullout and matrix failure along a single horizontal plane (Fig. 10a). High adhesion results in a single, clean fracture plane (Fig. 10c); intermediate adhesion exhibits components of both (Fig. 10b). Because the

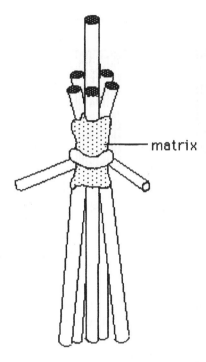
matrix

FIG. 9 Single-fiber pullout from a microcomposite test. (From Ref. 22.)

ineffective length is reduced as the quality of the interface increases, the ultimate tensile strength of the composite increases as the IFSS is increased.

2. Compression Tests

The results of compression tests (e.g., ASTM D3410-75) also can lead to a qualitative evaluation of fiber–matrix adhesion. Samples with low interfacial strength

TABLE 2 Statistics for IFSS for the SFPOM and Microbond Test Using *para*-Aramid Fibers in Epoxy

Test method	Mean IFSS (MPa)	CV (%)
SFPOM	26.9	31.4
Microbond	32.1	19.6

Source: Ref. 22.

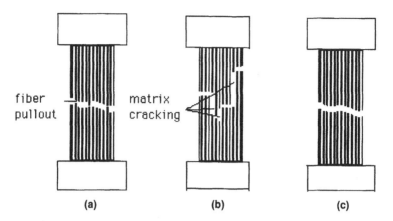

FIG. 10   Mode of tensile failure in a laminate with (a) poor IFSS, (b) moderate IFSS, and (c) high IFSS.

will exhibit fiber buckling and delamination, illustrated in Fig. 11a. Intermediate strength results in massive fiber buckling along the plane of maximum shear stress (Fig. 11b). Fiber compressive failure across several horizontal planes is an indication of a strong interface and high interfacial strength.

IV. SUMMARY

There exist many options to determine either the interface strength or the efficacy of the adhesion between the reinforcing fibers and the matrix in composite materials. There are a number of methods—both single- and multiple-fiber techniques—to directly measure the strength of the interfacial bond. The most common methods used are fiber pullout, microbond, fragmentation, and microindentation. These methods, although relatively simple to perform (especially in the case of fiber pullout), require a bit of experience in the sample preparation technique. Some of the tests (e.g., single-fiber pullout, fragmentation, and microbond) generally do not give a composite *in situ* value for the interfacial strength because of the essentially zero V_f. Other methods (e.g., fragmentation and microindentation) are not appropriate for all types of fibers because they rely on the fiber's either breaking cleanly (fragmentation) or not fibrillating (microintentation).

Measuring *in situ* interface strength is a bit more difficult. Both microindentation and SFPOM focus on individual fibers. As noted earlier, microindentation has a materials limitation as well as the need for a well-trained operator and precision equipment. SFPOM requires a degree of skill in sample preparation,

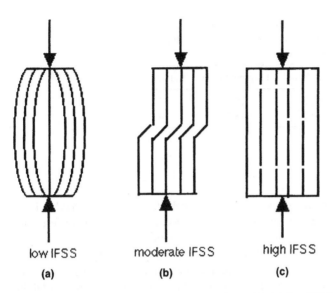

low IFSS   moderate IFSS   high IFSS

(a)   (b)   (c)

FIG. 11 Failure of a laminate in compression with (a) low IFSS, (b) moderate IFSS, and (c) high IFSS.

and the final V_f is not easily controlled. Laminates may be used to measure IFSS but only in a qualitative sense, as the efficacy of the interface can be observed from the test results, whereas the actual IFSS values are not measurable.

One final difficulty is that the different tests, even when performed in the same laboratory, often give different values for IFSS. This was best illustrated in a report by Pitkethly et al. [23] of a round-robin study conducted in February 1990. Twelve laboratories were given the task of measuring IFSS using the same carbon fiber and epoxy system. The laboratories were allowed to use whatever technique(s) [fiber pullout (single fiber and/or microbond), fragmentation, and/or microindentation] each felt appropriate. Table 3 contains collated data from all 12 laboratories for the different techniques applied to untreated carbon fibers (XAS,

TABLE 3 Collated Round-Robin Data for IFSS Using Untreated Carbon Fibers

	Pullout	Microbond	Fragmentation	Indentation
IFSS (τ) (MPa)	64.6	48.3	23.8	47.8
CV (%)	13	29	28	1

Source: Ref. 23.

TABLE 4 Calculated IFSS from Microbond,
Fragmentation, and Indentation Tests from "Laboratory 6"
Using Untreated Carbon Fibers

	Microbond	Fragmentation	Indentation
IFSS (τ) (MPa)	42.3	25.2	47.4
CV (%)	4.2	18.7	14.4

Source: Ref. 23.

Courtalds); all reported values of IFSS were calculated using the Kelly–Tyson equation. Not only is there variability in the measured IFSS for the different test methods, but there is also quite a bit of scatter within each method.

One of the 12 laboratories, identified by Pitkethly et al. [23] as "Laboratory 6," conducted microbond, indentation, and fragmentation tests using the same system, and these results are presented in Table 4. In this case, the variability within methods is greatly reduced, but there is still a difference in the measured value of IFSS; indentation gave an IFSS 188% greater than fragmentation and 112% greater than the microbond.

Work continues to refine the various tests to measure the strength of the interface and to interpret the results. Recently, researchers have questioned whether the average interfacial strength is accurate at all. Scheer and Nairn [24] argued that it is more appropriate, as well as accurate, to consider the critical energy release rate for interfacial crack growth as the determiner of failure load of microbond specimens. They note that others [25,26] have shown that the fracture process is seldom one of pure interfacial failure.

Ultimately, the choice of the appropriate tests and the interpretation of their results will lead to a better understanding of the fundamentals of adhesion in composites. This, in turn, will lead to better predictions of material properties and service lifetime in applications.

REFERENCES

1. L. T. Drzal, M. J. Rich, and P. F. Lloyd. Adhesion *16*:1 (1982).
2. H. F. Wu, D. W. Dwight, and N. T. Huff. Compos. Sci. Technol. *57*:975 (1997).
3. T. Morii, N. Ikuta, K. Kiyosumi, and H. Hamada. Compos. Sci. Technol. *57*:985 (1997).
4. T. Stern, A. Teishev, and G. Marom. Compos. Sci. Technol. *57*:1009 (1997).
5. E. Mäder. Compos. Sci. Technol. *57*:1077 (1997).
6. A. Kelly and W. R. Tyson. J. Mech. Phys. Solids *13*:329 (1965).
7. P. J. Herrera-Franco and L. T. Drzal. Composites *23*:2 (1992).

8. L. J. Broutman, in *Interfaces in Composites*, ASTM STP 452, ASTM, Philadelphia, 1969, pp. 27–41.

9. Z.-F. Li and A. N. Netravali. J. Appl. Polym. Sci. *44*:336 (1992).

10. J. Banbaji. Compos. Sci. Technol. *32*:195 (1988).

11. R. J. Gray. J. Mater. Sci. *19*:861 (1984).

12. M. R. Piggot. Compos. Sci. Technol. *57*:965 (1997).

13. B. Miller, P. Muri, and L. Rebenfeld. Compos. Sci. Technol. *28*:17 (1997).

14. S. Feih and P. Schwartz. Adv. Compos. Lett. *6*:99 (1997).

15. H. F. Wu and C. M. Claypool. J. Mater. Sci. Lett. *10*:269 (1991).

16. B. Fiedler and K. Schulte. Compos. Sci. Technol. *57*:859 (1997).

17. H. D. Wagner, S. L. Phoenix, and P. Schwartz. J. Compos. Mater. *18*:312 (1984).

18. F. T. S. Peirce. J. Textile Inst. *17*:T355 (1926).

19. A. N. Netravali, in *Handbook of Advanced Materials Testing* (N. P. Cheremisinoff and P. N. Cheremisinoff, eds.), Marcel Dekker, New York, 1994, p. 313.

20. A. N. Netravali, R. B. Henstenburg, S. L. Phoenix, and P. Schwartz. Polym. Compos. *10*:226 (1989).

21. G. Kalinka, A. Leistner, and A. Hampe. Compos. Sci. Technol. *57*:845 (1997).

22. Y. Qiu and P. Schwartz, J. Adhesion Sci. Technol. *5*:741 (1991).

23. M. J. Pitkethly, J. P. Favre, U. Gaur, J. Jakubowski, S. F. Mudrich, D. L. Caldwell, L. T. Drzal, M. Nardin, H. D. Wagner, L. Di Landro, A. Hampe, J. P. Armistead, M. Desaeger, and I. Verpoest. Compos. Sci. Technol. *48*:205 (1993).

24. R. J. Scheer and J. Nairn. J. Adhesion *53*:45 (1995).

25. C. T. Chou, U. Gaur, and B. Miller. Compos. Sci. Technol. *48*:307 (1993).

26. C. T. Chou, U. Gaur, and B. Miller. J. Adhesion *40*:245 (1993).

10

The Effect of Fiber Surface on the Thermal Properties of Fibrous Composites

YASSER A. GOWAYED Department of Textile Engineering, Auburn University, Auburn, Alabama

I. INTRODUCTION

Fibrous composites are made by infiltrating a fibrous assembly with a matrix material (e.g., polymer, ceramic). The matrix fills the space between fibers and bonds to fiber surfaces, creating a fiber–matrix interface. The properties of this interface depend on the fiber surface physical and chemical characteristics, the matrix material type, and the composite processing temperature [1–3].

Fibrous assemblies could be manufactured from short fibers (e.g., nonwovens) or long continuous fibers with various spatial arrangements (e.g., parallel, woven, braided). Although most of the fibers used in manufacturing composite materials are ceramic fibers (e.g., glass, carbon, SiC), other fiber types are also used, such as Kevlar® and steel.

The role of matrix and fibers in a given composite depend on their material type. Ceramic materials and metals have high stiffness, low thermal expansion coefficient, and high thermal conductivity when compared to polymers. By combining two or more of these material types as fibers and matrix, we can engineer composite materials with anisotropic properties that could not be achieved using a single type material.

The fiber–matrix interface has a significant effect on the composite thermal properties. Perfect bonding between fibers and matrix is typically desired; but interfacial debonding and boundary resistance could be found and may arise from disparity in the properties (e.g., bulk density, chemical structure) of the two materials meeting at the interface. Mismatches in the coefficient of thermal expansion (CTE) of the fiber and the matrix can cause incomplete thermal contact at the interface. On cooling from the fabrication temperature, the CTE mismatch can result in an interfacial gap, especially in those composites with poor or no adherence between the fibers and matrix. Another source of interfacial resistance is the physical irregularities at the boundaries. Surface roughness or the presence of impurities (e.g., oxide films, bonding agents) can result in a poor contact between the phases [4].

Analytical studies have shown that the thermal conductivity of a composite depends on the value of the thermal conductivity of the fibers and the matrix, the fiber volume fraction, and the distribution of the fibers within the composite (e.g., lamina, woven, braided). A number of recent studies also identified the critical role of an interfacial thermal barrier resistance at the fiber–matrix interface on the effective thermal conductivity of composites [5–7].

When a composite material is subjected to heat, both of its constituents (fibers and matrix) exhibit an expansion or a contraction. The fiber–matrix interface also contributes to this thermal deformation. Poor contact between fibers and matrix can lead to a weak composite and affect the composite overall CTE.

In this chapter, the effect of fiber interface and fiber coating on the thermal behavior of composite materials will be discussed. Two thermal properties of the composite will be studied: (1) thermal conductivity and (2) thermal expansion coefficients. The discussion will focus on composites made from parallel fibers (i.e., unidirectional lamina). This will allow us to separate the effect of fiber interface from the effect of fiber arrangement on the thermal properties parallel and transverse to the fiber direction.

II. EFFECT OF FIBER INTERFACE ON THERMAL CONDUCTIVITY OF FIBROUS COMPOSITES

Heat conduction in one dimension for an isotropic material is stated mathematically by Fourier Law as follows:

$$q = -KA \frac{dT}{dx} \tag{1}$$

where q is heat flow, K is thermal conductivity, T is temperature, and x is the direction of heat flow. The minus sign is a consequence of the second law of thermodynamics, which requires that heat flow in the direction of the lower temperature. If the rate of heat transfer is in watts, the area is in square meters, the temperature is in degrees centigrade, and x is in meters, then the thermal conductivity k will have the units of watts per meter per degree centigrade (W/m°C).

For anisotropic materials, such as fibrous composites, the relationship is different than that presented in Eq. (1). In this section, the thermal conductivity of a unidirectional composite with fibers laid parallel to each other in a lamina form is presented.

For the unidirectional composites shown in Fig. 1, we can assume that the heat conduction parallel to the fiber direction will only involve planar isotherms. This assumption is only valid for steady-state heat flow. However, for transient heat flow, transverse heat flow between the fibers and matrix is expected to occur, especially for components with significantly different thermal conductivities. If the thickness of the interface region is negligible, the longitudinal thermal conductivity of the composite will only be affected by the fibers and matrix thermal conductivities and will be unaffected by the fiber–matrix interfaces as follows:

$$K_l = K_{fl} V_f + k_m V_m \tag{2}$$

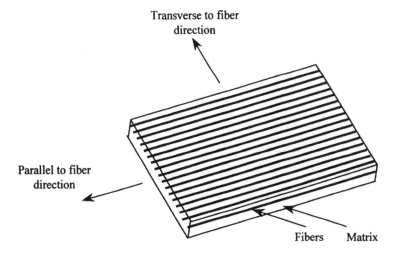

FIG. 1 Unidirectional composites.

where K_l is the thermal conductivity of the composite in the direction parallel to the fiber direction, K_{fl} and K_m are thermal conductivity of the fibers and matrix, respectively, and V_f and V_m are volume fraction of the fibers and matrix, respectively.

In the direction perpendicular to the fibers' direction, the interface plays an important role in the heat transfer operation. Interfacial heat transfer across the fiber–matrix interface, as part of the overall heat transfer through the composite, is the sum of the solid conduction through points or areas of direct contact between fibers and matrix, the gaseous conduction through the gas phase present within the interface at areas of debonding, and the radiation through the gaps. It was found [8] that the value of the radiative conductance is negligible.

The solid conduction through areas of contact between fibers and matrix could be analyzed by utilizing the cylindrical model shown in Fig. 2. A fiber, with radius a and thermal conductivity K_{ft} in the radial direction, is embedded in a cylinder of matrix with radius b and thermal conductivity K_m. The fiber has an interface thermal barrier of h_f.

The first step in determining the thermal conductivity of the composite in the direction transverse to the fiber direction is to derive an expression for the temperature distribution under the steady-state heat transfer conditions. This was

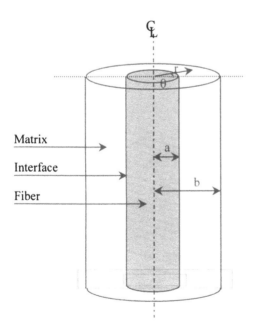

FIG. 2 Cylindrical model used to calculate transverse thermal conductivity.

given in Ref. 9 for a circular cylinder dispersion (e.g., fiber) oriented perpendicular to the heat flow as

$$T_f = rA \cos \theta \tag{3a}$$

$$T_m = (\nabla T)r \cos \theta + (B/r) \cos \theta \tag{3b}$$

where T_f is the temperature distribution inside the cylindrical dispersion, T_m is the temperature distribution in the surrounding matrix, r and θ represent the two-dimensional polar coordinate system, ∇T is the temperature gradient along the radial direction, and A and B are constants to be solved using the following boundary conditions:

$$K_m \left(\frac{\partial T_m}{\partial_r} \right) = K_{fr} \left(\frac{\partial T_f}{\partial r} \right) \quad \text{at } r = a \tag{4a}$$

$$K_{fr} \left(\frac{\partial T_f}{\partial r} \right) = h_c(T_m - T_f) \quad \text{at } r = b \tag{4b}$$

Notice that in the absence of an interfacial thermal barrier (i.e., $h_c = \infty$), Eq. (4b) becomes

$$T_m = T_f \tag{4c}$$

Substituting the temperature distribution equations in the boundary condition equations yields the values for the constants A and B. From Eqs. (4a)–(4c), the thermal conductivity for the composite in the transverse direction to the fibers could be derived as [10]

$$K_{comp} = K_m \frac{\left[\left(\dfrac{K_{ft}}{K_m} - 1 - \dfrac{K_{ft}}{ah_c} \right) V_f + \left(1 + \dfrac{K_{ft}}{ah_c} + \dfrac{K_{ft}}{K_m} \right) \right]}{\left[\left(1 + \dfrac{K_{ft}}{ah_c} - \dfrac{K_{ft}}{K_m} \right) V_f + \left(1 + \dfrac{K_{ft}}{ah_c} + \dfrac{K_{ft}}{K_m} \right) \right]^{-1}} \tag{5}$$

where V_f is the fiber volume fraction ($V_f = a^2/b^2$ for the cylinder shown in Fig. 2). Again, in the absence of an interfacial thermal barrier (i.e., $h_c = \infty$), Eq. (5) becomes

$$K_{comp} = K_m \frac{\left[\left(\dfrac{K_{ft}}{K_m} - 1 \right) V_f + \left(1 + \dfrac{K_{ft}}{K_m} \right) \right]}{\left[\left(1 - \dfrac{K_{ft}}{K_m} \right) V_f + \left(1 + \dfrac{K_{ft}}{K_m} \right) \right]^{-1}} \tag{6}$$

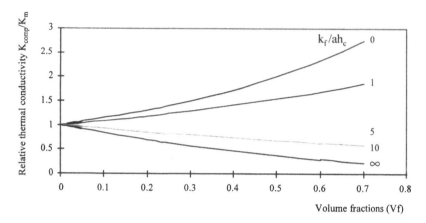

FIG. 3   The relative thermal conductivity of unidirectional composite materials for $K_f/K_m = 5$ utilizing the cylindrical model for a range of values of K_f/ah_c.

Equation (6) is in agreement with the solution presented by Rayleigh [11]. Notice that Eqs. (5) and (6) did not consider the interaction between fibers. Accordingly, these equations are accurate only for a low-fiber-volume fraction.

As a numerical example, Figure 3 shows the relative thermal conductivity of unidirectional composite materials for $K_f/K_m = 5$ (i.e., E-glass/epoxy composite) utilizing the cylindrical model for a range of values for K_f/ah_c from 0 (i.e., complete thermal barrier, $h_c = \infty$) to ∞ (i.e., no thermal barrier, $h_c = 0$). In the case of $K_f/ah_c = 0$, the composite effective thermal conductivity corresponds to predictions of thermal conductivity of a composite with perfect interface. In the case of $K_f/ah_c = \infty$, the composite effective thermal conductivity corresponds to the thermal conductivity of the matrix with voids (i.e., the fibers inside the matrix are replaced by voids). This happens because the thermal barrier at the fiber–matrix interface is very high and does not allow any heat transfer.

III. EFFECT OF FIBER COATING ON THE THERMAL CONDUCTIVITY OF COMPOSITES

In many cases, the coating of fibers is essential for fiber productivity and usability. For example, coating of glass fibers during production is crucial to avoid generating Griffith cracks that could drastically reduce the strength of the fibers. Also, coating carbon fibers with a polymer layer allows assembling the yarns into fabrics by weaving or braiding.

The difference between the fiber–matrix interface behavior versus the behavior of coated fibers is influenced by the size of the coating layer. We can assume that the first condition is a special case of the latter. The coating layer introduces two-interface thermal barriers: one layer between the fiber and the coating and the other between the coating and the matrix.

The longitudinal thermal conductivity for unidirectional composites (see Fig. 1), under steady-state heat transfer conditions, is not affected by the interfacial thermal barrier and can be expressed as

$$K_l = K_{fl}V_f + k_mV_m + K_cV_c \tag{7}$$

where K_l is the thermal conductivity of the composite in the direction parallel to the fiber direction, K_{fl}, K_m, and K_c are the thermal conductivities of the fibers, matrix, and coating, respectively, and V_f, V_m, and V_c are the volume fraction of the fibers, matrix, and coating, respectively.

In the transverse direction, the same procedure used in the previous section is used to derive the thermal conductivity of the composite (refer to Fig. 4). Substitution of the temperature-distribution equations for the fibers, matrix, and coating in the boundary condition equations yields the equation for the transverse thermal conductivity [12]:

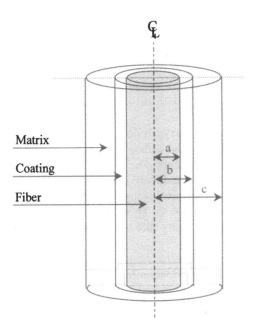

FIG. 4 Cylindrical model used to calculate the transverse thermal conductivity of the coated fiber composites.

$$K_{\text{comp}} = K_m \frac{\left[1 - e_m\left(\dfrac{V_f}{a^2} + \dfrac{V_c}{b^2 - a^2}\right)\right]}{\left[1 + e_m\left(\dfrac{V_f}{a^2} + \dfrac{V_c}{b^2 - a^2}\right)\right]^{-1}} \tag{8}$$

where

$$\frac{e_m}{b^2(\nabla T)} = \frac{1 + 2\left[a^2\left(\dfrac{K_c - K_f}{K_c K_f} + \dfrac{1}{ah_{fc}}\right) - b^2\left(\dfrac{K_c + K_f}{K_c K_f} + \dfrac{1}{ah_{fc}}\right)\right]}{\left\{K_m\left[a^2\left(\dfrac{K_c - K_f}{K_c K_f} + \dfrac{1}{ah_{fc}}\right)\left(\dfrac{K_m - K_c}{K_c K_m} - \dfrac{1}{bh_{cm}}\right) + b^2\left(\dfrac{K_c + K_f}{K_c K_f} + \dfrac{1}{ah_{fc}}\right)\left(\dfrac{K_c + K_m}{K_c K_m} + \dfrac{1}{bh_{cm}}\right)\right]\right\}^{-1}}$$

h_{cf} and h_{cm} are the interfacial thermal barrier between fibers and coating and matrix, respectively, and ∇T is the temperature gradient in the radial direction. Because the effect of coated fibers on each other is again neglected, the above expression for thermal conductivity could only be applied for low-fiber-volume fractions.

Very few experiments were conducted to evaluate the effect of the fiber interface on the thermal conductivity of composites. In this section, the effect of carbon coating on silicon carbide fibers embedded in a reaction-bonded silicon nitride matrix is presented [6,7,13]. The composite was manufactured from a reaction-bonded SiN matrix reinforced with unidirectional SCS-6 SiC monofilaments. The fibers diameter was 142 μm and contained a carbon central core with a diameter of 37μm. The volume fraction of the fibers was 31% and the matrix had 31% voids with a mean pore size of 0.02 μm. The composite density was 2.17 g/cm³.

Some composite samples were subjected to oxygen flow at 600°C for 100 h to cause oxidation. This resulted in oxidizing the carbon coating on the fibers and some oxidation of the carbon core of the fibers. The flash diffusivity technique was used to measure the effective thermal diffusivity of the original and oxidized composite. The specific heat was measured using differential scanning calorimetry (DSC) from room temperature to 600°C. Thermal conductivity was calculated from the experimental thermal diffusivity data and the product of density and specific heat.

It was observed from the experimental data, for samples tested in helium, nitrogen, and vacuum environments, that the oxidation treatment has no effect on thermal diffusivity and heat conduction in the direction parallel to the fibers. Thermal conductivity values of untreated and oxidized composite for different testing temperatures in all environments were almost identical.

For heat flow transverse to the fiber direction, the thermal diffusivity/conductivity is strongly affected by the existence of the interfacial thermal barrier. The gas phase (i.e., vacuum, nitrogen, and helium) entrapped inside the interface region contributes to the total interfacial thermal conductance of the composite. Removal of the interfacial carbon layer by oxidation lowered the conductance of the composite. The measurement in vacuum, in which the contribution of the gaseous conductance is suppressed, indicated that the contact conductance increases rapidly with temperature. Radiative conductance was found to be negligible.

IV. EFFECT OF FIBER COATING ON THE THERMAL EXPANSION COEFFICIENT OF COMPOSITES

In this section the CTE of unidirectional composite with fibers laid parallel to each other in a lamina form, as shown in Fig. 1, is discussed. The stress-strain behavior can be represented, in tensor form, as follows:

$$\varepsilon_{ij} = S_{ijkl}\sigma_{kl} + \alpha_{ij}\Delta T$$
$$\sigma_{ij} = C_{ijkl}\varepsilon_{kl} - \Gamma_{ij}\Delta T$$

where σ is the stress vector, ε is the strain vector, C is the stiffness tensor, S is the compliance tensor ($S = C^{-1}$), α is the vector of CTE, Γ is the vector of thermal stresses ($\Gamma = C_{ijkl}\alpha_{kl}$), and ΔT is the temperature increment. These equations show that the internal stresses are dependent on the mechanical and thermal external stresses as well as the material properties.

For fibrous composites, the coefficient of thermal expansion is a function of the volume fraction of constituent materials, their CTEs, their relative stiffness, and the interface characteristics between the constituents. Fiber coating could have a large effect on the CTE of composites.

In polymer composites, fibers have a higher stiffness than the matrix; that is, E-glass fibers have an average elastic modulus of 72 GPa, whereas epoxy resins have an average modulus of around 5 GPa. This implies that fibers have a much lower CTE than that of the matrix (e.g., CTE for E-glass is $5 \times 10^{-6}°C^{-1}$, whereas the CTE for epoxy resins is around $45 \times 10^{-6}°C^{-1}$). For unidirectional polymer composites, the CTE in the direction parallel to the fibers is typically controlled by the fibers. In the direction transverse to the fiber direction, the matrix is free to move and is not constrained by the low CTE of the fiber.

Many models have been developed to understand and calculate the CTE of unidirectional fibrous composite [14–19] in the longitudinal and the transverse directions. Most of these models do not take into consideration the effect of the fiber–matrix interaction or fiber coating on the properties of the composite. Also, extension of these models to include the effect of voids and three-phase composites (i.e., fiber–coating–matrix) is not possible.

In this section, a new approach is presented and derived from basic principles of stresses and strains for anisotropic materials. Although this approach was derived for two- and three-phase materials, it could be easily extended to model n-phase materials with different internal structures and inclusions/voids.

Basic assumptions for this approach could be summarized as follows: (1) steady-state heat flow, (2) temperatures of fiber, coating, and matrix are constant and equal, and (3) all interfaces are perfectly bonded.

In a unidirectional composite, if we treat the fiber as a transversely isotropic material and the matrix and the coating as isotropic materials, the basic equations of stress, strain, and displacement, in polar coordinates, can be represented as follows (refer to Fig. 4) [20]:

$$u_{ir} = A_{2i} \frac{1}{r} \int_0^r T_i r \, dr + C_{1i} r + \frac{C_{2i}}{r} - \frac{v_i C_{3i} r}{E_i} \tag{9a}$$

$$\sigma_{ir} = -A_{1i} \frac{1}{r^2} \int_0^r T_i r \, dr + A_{3i} C_{1i} - \frac{A_{4i} C_{2i}}{r^2} \tag{9b}$$

$$\sigma_{i\theta} = A_{1i} \frac{1}{r^2} \int_0^r T_i r \, dr - A_{1i} T_i + A_{3i} C_{1i} + \frac{A_{4i} C_{2i}}{r^2} \tag{9c}$$

$$\sigma_{iz} = -A_{1i} T_i + 2v_i A_{3i} C_{1i} = -C_{3i} \tag{9d}$$

$$\varepsilon_{ir} = \frac{du_{ir}}{dr} \tag{9e}$$

$$\varepsilon_{i\theta} = \frac{u_{ir}}{r} \tag{9f}$$

$$\varepsilon_{iz} = \frac{C_{3i}}{E_i} \tag{9g}$$

where i refers to fiber, matrix, or coating; $0 \leq r \leq c$, u_{ir} is radial displacement; σ_{ir}, $\sigma_{i\theta}$, and σ_{iz} are stresses in radial, tangential, and axial directions, respectively; ε_{ir}, $\varepsilon_{i\theta}$, and ε_{iz} are strains in the r, θ, and z directions, respectively; α_{ia} and α_{it} are axial and transverse thermal expansion coefficients, respectively; E_i and v_i are the axial Young's modulus and Poisson's ratio, respectively; C_{1i}, C_{2i}, and C_{3i} are constants that need to be solved; and

$$A_{1i} = \frac{(\alpha_t^i + v_i \alpha_a^i) E_i}{1 - v_i^2}, \qquad A_{2i} = \frac{\alpha_t^i + v_i \alpha_a^i}{1 - v_i},$$

$$A_{3i} = \frac{E_i}{(1 + v_i)(1 - 2v_i)}, \qquad A_{4i} = \frac{(\alpha_t^i v_i + \alpha_a^i) E_i}{1 - v_i^2}$$

Notice that due to the symmetry about the axes and the uniformity in the axial direction of this fiber–coating–matrix system, all shear strains and stresses are zero.

To calculate the axial thermal expansion coefficient, where the fibers will control the composite behavior (especially in polymer composites), we assume the boundary conditions to be as follows:

$$u_{fr} = 0 \quad \text{at } r = 0 \tag{10a}$$

$$\sigma_{fr} = \sigma_{cr} \quad \text{at } r = a \tag{10b}$$

$$\sigma_{mr} = \sigma_{cr} \quad \text{at } r = b \tag{10c}$$

$$\sigma_{mr} = 0 \quad \text{at } r = c \tag{10d}$$

$$\varepsilon_{fz} = \varepsilon_{cz} = \varepsilon_{mz} \tag{10e}$$

Solving the previous equations for a specific material configuration such as fiber–interface–matrix or fiber–matrix) under the specified boundary conditions, we can obtain the constants C_{1i}, C_{2i}, and C_{3i}, and calculate the stresses and strains in Eqs. (9). Notice that the number of the constituents of the composite will determine the size of the problem. For example, for a two-phase composite (i.e., fiber and matrix only), the number of constants to be solved are six with three boundary conditions, whereas for a three-phase composite (i.e., fiber–coating–matrix), the number of constants is nine with five boundary conditions. If we define the longitudinal thermal expansion coefficient of a composite as

$$\alpha_a = \frac{\varepsilon_{iz}}{T_b}$$

where ε_{iz} is the strain of the fiber, coating, or matrix [refer to Eq. (10e)] in the axial direction, then we can calculate the thermal expansion coefficient of the composite in the axial direction.

To calculate the transverse thermal expansion coefficient, where the matrix will control the composite behavior, we assume the boundary conditions to be as follows:

$$u_{fr} = 0 \quad \text{at } r = 0 \tag{11a}$$

$$\sigma_{fr} = \sigma_{cr} \quad \text{at } r = a \tag{11b}$$

$$u_{fr} = u_{cr} \quad \text{at } r = a \tag{11c}$$

$$\sigma_{mr} = \sigma_{cr} \quad \text{at } r = b \tag{11d}$$

$$u_{mr} = u_{cr} \quad \text{at } r = b \tag{11e}$$

$$\sigma_{mr} = 0 \quad \text{at } r = c \tag{11f}$$

Again, notice that the number of the constituents of the composite will determine the size of the problem. For a two-phase composite, the number of constants to be solved are six with four boundary conditions, whereas for a three-phase composite, the number of constants is nine with six boundary conditions.

The transverse thermal expansion coefficient of the composite could be defined as the value of the strain at $r = c$ (the outer surface of the composite cylinder):

$$\alpha_t = \frac{\varepsilon_{mr}}{T_c}$$

where ε_{mr} is the strain of the matrix in the radial direction. Using this equation, we can calculate the thermal expansion coefficient of the composite in the transverse direction.

The effect of the CTE of the coating on E-glass fibers–thermoplastic coating–epoxy matrix composites was studied using the above-mentioned model. In this analysis, the E-glass elastic modulus was taken as 72 GPa with a Poisson's ratio of 0.25 and a CTE of 5×10^{-6} °C^{-1} in longitudinal and transverse directions. Elastic modulus for the coating material was taken as 0.2 GPa with a Poisson's ratio of 0.4 and a variable CTE. For the epoxy matrix material, the elastic modulus was taken as 4.9 GPa with a Poisson's ratio of 0.35 and CTE of 45×10^{-6} °C^{-1}. The fiber volume fraction was taken as 34.6%, the coating volume fraction as 7.3%, and the matrix volume fraction as 58.1% (i.e., from Fig. 4, $a = 1$, $b = 1.1$, and $c = 1.7$).

It can be shown from Fig. 5 that the longitudinal CTE of the composite was not affected by the change in the CTE of the coating. This is an expected result

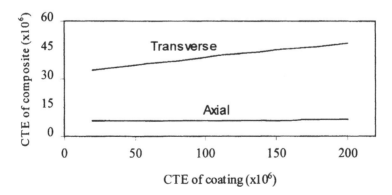

FIG. 5 The effect of the CTE of coating on the longitudinal and transverse CTEs of unidirectional E-glass fiber–thermoplastic coating–epoxy matrix composites.

because the elastic modulus of the coating is much lower than that of the fiber. The fiber dominated the strain and, accordingly, the CTE behavior of the composite. In the transverse direction, the effect of the value of the CTE of the coating on the CTE of the composite is dramatic. A linear increase of the value of CTE could be observed with the increase in the value of the CTE of the coating. We can conclude from this graph that the effect of fiber coating, in unidirectional polymer composites, is more evident in the transverse direction than in the longitudinal direction.

The effect of the coating thickness on the longitudinal and transverse CTE of unidirectional E-glass–thermoplastic coating–epoxy matrix composite was also investigated. In this analysis, the same properties used in the previous example were used. The CTE of the coating was fixed at $100 \times 10^{-6}\,°C^{-1}$ and the thickness was varied from 10% of the fiber thickness to 30% of the fiber thickness.

It can be seen from Fig. 6 that the longitudinal CTE of the composite is still controlled by the CTE of the fiber. The coating has almost no effect on the longitudinal behavior of the composite. Again, this could be attributed to the low stiffness of the thermoplastic coating when compared to the stiffness of the ceramic E-glass fiber. In the transverse direction, it can be concluded from Fig. 6 that the thicker the fiber coating, the higher the CTE of the composite in the transverse direction.

It could be observed from Figs. 5 and 6 that the effect of fiber coating on the CTE of composites is constrained by its relative stiffness when compared to the fiber and matrix properties. For unidirectional composites in a direction parallel to the fiber direction, if the coating has low stiffness when compared to the fiber and the matrix, the CTE of the composite will not be affected by the coating

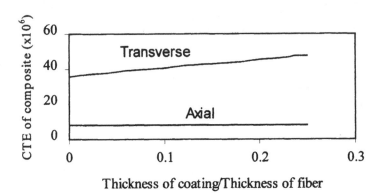

FIG. 6 Effect of thermoplastic coating thickness on the longitudinal and transverse CTEs of unidirectional E-glass fiber–thermoplastic coating–epoxy matrix composites.

properties. If the coating has stiffness comparable to that of the matrix and the fiber, its CTE will contribute to the longitudinal CTE of the composite.

In the transverse direction, if the stiffness of the coating is lower than that of the matrix, the coating will cause the CTE of the composite to increase. If the situation is reversed, the stiffness of the coating is higher than that of the matrix, the CTE of the unidirectional composite in the transverse direction will decrease.

Finally, we can state that the contribution of the fiber coating to the CTE of unidirectional composites is not only dictated by the CTE of the coating and its thickness, but also by the stiffness of the coating with respect to the fiber and matrix.

REFERENCES

1. G. Krekel, K. Huttinger, W. Hoffman, and D. Silver. J. Mater. Sci. 29:2968 (1994).
2. G. Krekel, K. Huttinger, and K. Hoffman. J. Mater. Sci. 29:3461 (1994).
3. G. Krekel, U. Zielke, K. Huttinger, and K. Hoffman. J. Mater. Sci. 29:3984 (1994).
4. Y. Chiew and E. Glandt. Chem. Eng. Sci. 42(11):2677 (1987).
5. Y. Benveniste. J. Appl. Phys. 61(8) (1987).
6. H. Bhatt, K. Donaldson, D. Hasselman, and R. Bhatt. J. Am. Ceram. Soc. 75(2): 334 (1990).
7. H. Bhatt, K. Donaldson, D. Hasselman, and R. Bhatt, in *Thermal Conductivity 21* (C. J. Cremetrs and H. Fine, eds.), Plenum, New York, 1990.
8. W. Leung and A. Tam. J. Appl. Phys. 63(9) (1988).
9. Y. Beneveniste, T. Chen, and G. Dvorak. J. Appl. Phys. 67(6) (1990).
10. D. Hasselman and L. Johnson. J. Compos. Mater. 21:508 (1987).
11. Lord Rayleigh. Phil. Mag. 34:481 (1892).
12. Y. Lu, K. Donaldson, D. Hasselman, and J. Thomas. J. Compos. Mater. 29(13) (1995).
13. H. Bhatt, K. Donaldson, and H. Hasselman. J. Amer. Ceram. Soc. 73(2):312 (1990).
14. V. M. Levin. Mekhanika Tverdogo Tela 2(1):88 (1967) (in Russian).
15. R. Hill. J. Mech. Phys. Solids 12:199 (1964).
16. B. W. Rosen and Z. Hashin. Int. J. Eng. Sci. 8(2):157 (1970).
17. R. A. Schapery. J. Compos. Mater. 2(3):380 (1968).
18. B. W. Rosen. Proc. Royal Soc. London A 319:79 (1970).
19. C. Hsueh and P. F. Becher. J. Am. Ceram. Soc. 71:438 (1988).
20. S. P. Timoshenko and J. N. Goodier, *Theory of Elasticity*, McGraw-Hill, New York, 1934, pp. 433–484.

11
Design and Permeability Analysis of Porous Textile Composites Formed by Surface Encapsulation

MATTHEW DUNN   Fiber Architects, Maple Glen, Pennsylvania

I. INTRODUCTION

Textile structures have characteristic interstices (weaves), loops (knits), or voids (nonwovens) that can be utilized as pores. In a porous, rigidized structure, this could be realized by surface encapsulation of yarns with a hardened resin matrix. The permeability of the structure would then be controlled by the formed fabric structure. This is in contrast to the present reinforced composites, which are produced as fully infiltrated solid materials, wherein permeability is a defect. Creating selectively porous materials will require specialized materials and/or processing combined with imaginative structural design.

The materials described herein are porous composite materials formed from a textile reinforcement and a matrix that infiltrates only the yarn structure, not the "gaps" between the yarns. If textile-reinforced composites with uniform, continuous voids can be produced, many different filtration applications become available, including particle, fluid, and air and sound filtration. The well-developed field of textile processing could be combined with resin transfer molding or thermal sheet forming and utilized for composite production. Fabrics in excess of 3 m in width and continuous lengths can be manufactured, allowing a variety of large porous structures to be produced.

The formed composites were termed "renitent" composites, named to suggest the rigidity of the material without implying that the composites have porosity that deters from performance, but instead uses the engineered porosity as a performance goal. This terminology creates a unique name for a novel class of materials.

Composites were formed with woven, knitted, and nonwoven fabric reinforcement. The matrix was formed by resin transfer molding (RTM) of thermosetting epoxy or through coformed fabrics with a thermoplastic component utilized as the renitent composite matrix.

Permeability testing was performed on the renitent composites, and a model based on a comparison of Darcy's Law to MacGregor's textile description of the Kozeny–Carmen equation was created. This model provides a methodology to rank the permeability of materials based on the media porosity as determined by image analysis. The technique was found to correlate the permeability trend as found from the Darcy and MacGregor equations.

II. BACKGROUND

The term "textile composite" implies the use of textile fabrication techniques to produce the reinforcement in a composite system. This can refer to weaving, knitting, nonwovens, and braiding. A brief description of weaving, knitting, and nonwoven fabrics follows. Braiding is not addressed herein because no braided structures were used. A detailed description of braiding can be found in Ref. 1.

Classical weaving involves the interlacing of two orthogonal systems of yarns: the warp and filling. Each yarn system is manipulated so that the end fabric consists of parallel warp yarns and parallel filling yarns, with the warp and filling interlaced to provide stability (see Fig. 1). Yarn spacings dictate the size of the pore created between yarn intersections. This pore is termed the "interstice." From a planar standpoint, the perimeter of the interstice is dependent only on yarn spacing and final yarn width.

Knitted fabrics consist of yarns interlooped to form a stable continuous fabric. Knitted fabrics are formed in two families: weft and warp knitting. Weft knits are formed with one continuous yarn interlooped with itself (see Fig. 2). The loops formed create open planar space. The "loop" size depends on the final yarn diameter and the spacing of the courses and wales as created during fabrication.

Warp knitting consists of multiple yarns arranged in a parallel warp, with each yarn manipulated to interloop with its neighbors (see Fig. 3). Many different constructions are available, because each yarn can be made to interloop with one or multiple neighboring yarns in differing patterns. The more complex warp knit still has inherent porosity, as each pore is determined by the final yarn diameter and yarn spacings during construction.

Knitted fabrics exhibit much greater conformability than woven fabrics, and this may yield complex shape arrangements in rigid composites unobtainable with woven reinforcement [2]. A knit can conform to curved surfaces much easier

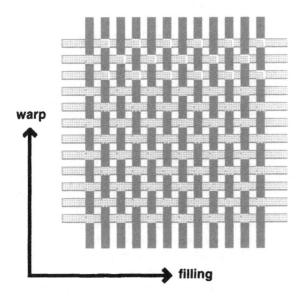

FIG. 1 Woven fabric construction.

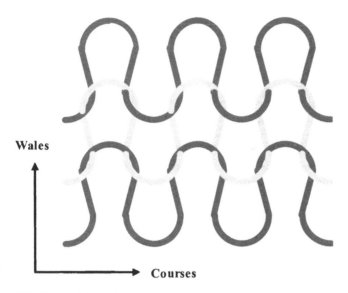

Wales

Courses

FIG. 2 Weft knitted fabric construction.

during layup, making production of spherical and other curved surfaces easier than with woven fabrics. Knits also provide multiple reinforcement directions and can be axially strengthened with weft insertion [3].

Nonwovens usually describe a fabric formed directly from fibers, but more generally describe a fabric that cannot be classified as woven, knitted, or braided. Although directional reinforcement is possible by using carded webs or fiber placement, generally nonwovens have a random fiber orientation (see Fig. 4). This random orientation is useful in composites when a closely isotropic material is desired. Nonwoven fabrics cannot achieve the same high-fiber-volume fractions as weaves and, consequently, have greater porosities than both woven and knitted fabrics.

The use of textile composites has been mainly limited to structural applications, with textile formation making layup and assembly of composites easier. Subsequently, the vast majority of their evaluation has been centered on their elastic and strength properties. Bogdanovich and Pastore [4] have an excellent review of these studies. A major advantage of textile composites is the ability to provide multiple reinforcement directions, greatly increasing the delamination resistance in multilayer structures [5]. This allows for the creation of thick composite panels with increased impact resistance when compared to laminates. Studies involving textile composites for nonstructural applications are limited at present.

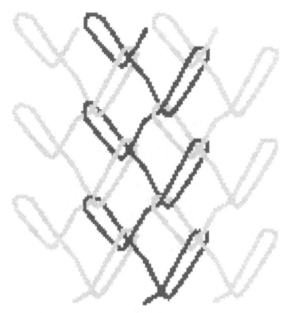

Basic Warp Knit Stitch

FIG. 3   Warp knitted fabric construction.

A.   Filtration and Flow

The filtration performance of fabrics have been examined for a number of years (e.g., see Refs. 6 and 7). Dickenson [8] classified the various applications of filtration media into four broad families:

1.   Solids–gases separation: mostly air filters, also including industrial gas processing, etc.

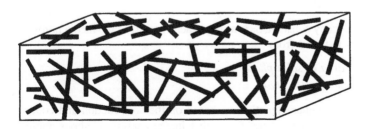

FIG. 4   Nonwoven fabric construction.

2. Solids–liquids separation: many mechanical filtration areas, differing in volume and size requirements
3. Liquids–liquids separation: especially phases having different densities and boiling points
4. Solids–solids separation: sieving of differing solid phases, especially those made of different particular sizes

The specific end uses of textiles for filtration are numerous and can fall anywhere within the range of textile markets. These include the geotextile (erosion control), medical (hemofiltration), household (air filters), and industrial (dust removal) textile markets, to name a few.

Filtration can be described mathematically in many ways. One approach is to describe the flow of air or fluid through a filtration media based on the properties of the air or fluid penetrant. Fluid flow will be used to describe both air and fluid flow henceforth, as air can be assumed to be a very low viscosity fluid. Fluid flow through a porous membrane can be described by the relationship

$$B = \frac{\mu t V}{\Delta P} \tag{1}$$

where μ is the viscosity of the fluid, t is the thickness of the membrane, V is the velocity of the fluid, and ΔP is the pressure drop across the membrane. Equation (1) is one form of Darcy's Law [9], named for the French engineer who published an equivalent relation based on experiments with the water supplies for the fountains of the city of Dijon [10]. Here, B is considered as a constant for the relationship between V and ΔP and is called Darcy's constant or (more commonly) the permeability of the membrane. The value for B will depend on the type of porous media and the pore geometry.

Following the reasoning that the void content in a porous media is a primary factor in the media permeability, the Kozeny–Carman equation was developed to provide a description of fluid flow based on the filtration media properties. One form of this equation [11] is

$$B = \frac{1}{K_0 S_0^2} \frac{\Phi^3}{(1 - \Phi)^2} \tag{2}$$

where K_0 is the Kozeny constant, S_0 is a shape factor, and Φ is the media porosity. The shape factor is found from

$$S_0 = \frac{Surface\ area\ of\ solid\ phase}{Volume\ of\ solid\ phase} \tag{3}$$

where the solid phase characteristics are based on the construction and content of the filtration media. It has been found that image analysis methods applied to Eq. (2) may yield reasonable results when predicting the Kozeny–Carman parameters [12]. Image analysis provides a realistic measurement of pore size and distribution.

The Kozeny–Carman equation is designed for a packed porous bed with random, tortuous paths available for fluid flow [11]. This assumption is representative of the pore assembly present in a nonwoven fabric with random fiber orientations throughout the fabric structure, but it has also been shown to have the capability of being extended to woven and knitted textile structures [13].

Another way of characterizing a filtration media is by the resistance the media presents to fluid flow. Pierce [14] attempted to describe the resistance a textile material will exhibit based on material properties:

$$R = \frac{S_0 S^2 \rho^2 (1 - \Phi)^2}{\Phi^3} \tag{4}$$

where R is the resistance to flow, S is the surface of void channels per unit mass (or the total specific surface of the media mass), and ρ is the overall mass density of constituents within the media. Pierce noted that this is, in fact, a method of "extreme simplification" and that "there is no pretense that the form assumed is geometrically similar to the form to be studied." He suggested that the shape factor be determined empirically, with the other variables being calculated from derived relationships. This equation was an early attempt at describing fluid flow specifically through textile materials.

Ergun [15], working with experimental results gathered from filtration media based on different shape parameters, developed the following relationship for filtration media with cylindrical solid constituents:

$$B = \frac{(\phi D)^2 \Phi^3 g_c}{150(1 - \Phi)^2} \tag{5}$$

where ϕ is a shape factor, D is the diameter of the cylinders, and g_c is the gravitational constant.

MacGregor [13] extended the Kozeny–Carman equation for a textile assembly in order to model the flow of dyes through textile yarn packages. If the solid phase is composed of circular fibers with diameter d and length, l, it can be easily found that for textile beds,

$$S_0 = \frac{4\pi d l}{\pi d^2 l} = \frac{4}{d} \tag{6}$$

By substitution, the Kozeny-Carman equation becomes

$$B = \frac{d^2}{16K_0} \frac{\Phi^3}{(1 - \Phi)^2} \tag{7}$$

The MacGregor equation provides an easy method to predict permeability based on fiber diameter and fabric porosity; or, the equation can be used to predict porosity values based on permeability measurements. This may prove to be useful, as a real porosity measurement may involve image analysis, a long and expensive process. Permeability measurements can be done rapidly by machine analysis.

Chiekhrouhou and Sigle [16] and Starr [17] are among a number who have attempted to relate fabric construction to pore geometry. The existing methods involve describing the yarn interactions and creating a unit cell for the fabric, with each unit cell exhibiting different porosity values. All of the unit cell models are based on woven cloth geometry.

Nonwoven filters can have much lower cost than woven or knitted filters, providing an economical and disposable filtration media for a number of applications. Nonwovens have been employed as filtration media for a number of years, in products such as tea bags, air filters, surgical dressings, and many others [18]. Their filtration performance has been greatly inspected, with most studies based on obtained particle filtration results. Van Den Brehel and De Jong [19] evaluated the Kozeny constant obtained from transverse flow experiments in textile products, taking the fiber type into consideration. Mahale et al. [20] matched the refractive index of a liquid penetrant to the fibrous mat under investigation, with mixed results.

B. Thermoplastic Composites

Various studies on the processing of thermoplastic matrix composite systems (e.g., Refs. 21–23) have been completed. All of the existing studies have been aimed at producing composites with (close to) zero porosity. The obstacles to efficient formation of thermoplastic composites include the inherently high molecular weights and accompanying high melt viscosity of the thermoplastic resin, requiring modified processes using extreme heat and/or pressure. Because stiffness, toughness, and a high melt temperature are requirements for most reinforced composite matrix materials, expensive resins such as poly(ether ether ketone) (PEEK) have been utilized. This has resulted in the impression that thermoplastic composites necessarily have high costs associated with them. If nonstructural applications for thermoplastic composites can be identified, then less expensive and widely available matrices (polyester and polyethylene, for example) could be used. Thermoplastic composites would then be more attractive, and further research into their processing would have a driver.

Limited studies have focused on the application of textile fabrics to reinforcement in inexpensive thermoplastic composite systems. Mayer et al. [24] suggested that knitted thermoplastic composites may yield inexpensive, formable parts for the automotive industry, but industry reaction has been slow to match expectations.

III. COMPOSITE FABRICATION

A. Designing for Porosity

In traditional textile-reinforced composite production, some measure of the fabric's fiber volume fraction is necessary to predict composite stiffness properties. The remaining volume of the fabric must be air; the air is then essentially replaced by resin. The result is a composite sheet composed of fiber (traditionally between 50% and 60%), resin matrix, and minimal (under 2%) air bubbles trapped by the resin. Any degree of air porosity over 2% is usually considered unacceptable [22].

The design of the fabrics used for this study was to keep the pore structure of the textile exposed, bounded by the placed yarn systems. More exactly, the yarns were encapsulated by resin (see Fig. 5), but the pores formed in the fabric formation process were kept open. Unless a yarn is a monofilament, there must be some volume within the total yarn cross section occupied by air, quantified by a yarn packing factor (φ). The encapsulation of the yarn is designed to replace the volume of air within the yarn with resin, hopefully forming an end structure with the same geometric three-dimensional structure as was fabricated. In theory, this step only stiffens the yarns into a rigidized fabric.

An untextured multifilament yarn usually has about 70% packing factor within the yarn or 30% air volume [4]. Therefore, the amount of resin used for encapsulation is nearly half that used for solid object production. Based on the input yarn properties, the amount of resin needed for encapsulation based on the packing factor can be found.

The area of fibers within the yarn cross section can be found from

$$A_f = \frac{N_d}{\rho(9 \times 10^5)} \tag{8}$$

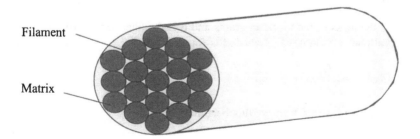

Filament

Matrix

FIG. 5 Filament yarn encapsulation.

where A_f is the total fiber area within the yarn cross section (cm^2), N_d is the yarn denier (denier is linear density of the yarn in g/9000 m), and ρ is the fiber density (g/cm^3). The total area occupied by the yarn in cross section (A_y) is then

$$A_y = \frac{N_d}{\varphi\rho(9 \times 10^5)} \tag{9}$$

where φ is the fiber packing factor within the yarn, or

$$\varphi = \frac{\text{Total fiber area}}{\text{Total yarn area}} \tag{10}$$

The area of the matrix needed for encapsulation (A_m) is then

$$A_m = A_y - A_f$$

By substitution,

$$A_m = \frac{N_d}{\varphi\rho(9 \times 10^5)} - \frac{N_d}{\rho(9 \times 10^5)} = \frac{(1 - \varphi)N_d}{\varphi\rho(9 \times 10^5)}$$

or

$$A_m = \frac{1 - \varphi}{\varphi} A_f \tag{11}$$

To determine the total volume of resin needed for encapsulation of a fabric, the total volume of fiber in the fabric must be known. This can be found from

$$\Omega_f = A_f l$$

where Ω_f is the volume of fiber in the fabric and l is the total length of yarn in the fabric. The total volume of fiber can be found using real measurements from

$$\Omega_f = \frac{m}{\rho} \tag{12}$$

where m is the mass of the fabric in grams and ρ is the fiber density (in g/cm^3).
The total matrix volume (Ω_m) needed is found from

$$\Omega_m = A_m l = A_m \frac{\Omega_f}{A_f}$$

By substitution and simplification, this becomes

$$\Omega_m = \frac{(1 - \varphi)m}{\varphi\rho} = \frac{1 - \varphi}{\varphi} \Omega_f \tag{13}$$

Equation (13) will yield the volume of resin needed in order to achieve full encapsulation of all yarns within the fabric undergoing impregnation.

B. Resin Transfer Molding

Resin transfer molding (RTM) is a process whereby catalyzed resin is transferred or injected into an enclosed mold in which the reinforcement has been placed. This is achieved by pulling a vacuum across the fabric reinforcement, simultaneously drawing resin through the mold and removing air from the porous structure (see Fig. 6). Traditional RTM places the mold under heat and pressure to catalyze resin curing.

A modified RTM method was used to produce the renitent composites. The predetermined needed amount of resin was worked into the fabric by hand, with some extra amount of resin added in case of overbleeding. Flow along a multifilament yarn has been shown to occur through capillary flow [25], whereby the fluid will pass along the yarn length, between the fiber filaments. Capillary flow will occur before the fluid flows into the pores between yarns, leaving the interstices between yarns open until capillary action has occurred.

In order to observe that capillary action will indeed occur, four sample fabrics were chosen to undergo RTM: a tight 1 × 1 rib weft knit, a loose satin warp knit, and two double weaves. The rib knit used nylon multifilament yarn; the loose warp knit was made of fiberglass; and the double weaves were formed with a cotton warp and Kevlar® 195 filling (see Table 1).

Figures 7–10 demonstrate that capillary flow was the preferred flow path for the infiltrating resin. Instead of filling the pores, the resin flowed along the yarn surfaces of the 1 × 1 rib (Fig. 7) even though it appears that too much resin was applied. The double cloths (Figs. 9 and 10) both show that even with two planes to flow across (face and back weave) and two fiber lengths (staple and filament), the resin fully infiltrated the yarns on the surface before penetrating into the woven interstices.

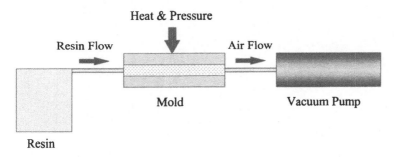

FIG. 6 Traditional RTM process.

TABLE 1 Renitent Composite Parameters, RTM Processing

Fabric construction	Yarn content	Fabric weight (oz/yd^2)
1 × 1 Rib knit	200 Denier nylon	125
Loose tricot warp knit	200 Denier glass	40
Weave 102, modified twill	Warp: 300 denier cotton; Fill: 195 denier Kevlar	260
Weave 201, face twill	Warp: 300 denier cotton; Fill: 195 denier Kevlar	260

C. Commingled Thermoplastic Composites

Commingled yarns, containing both reinforcement and matrix components during composite production, may be less expensive than producing renitent composites by RTM. The steps of resin mixing, application, and vacuuming are avoided. Instead, the formed fabric is placed directly into a heated mold to melt the thermo-

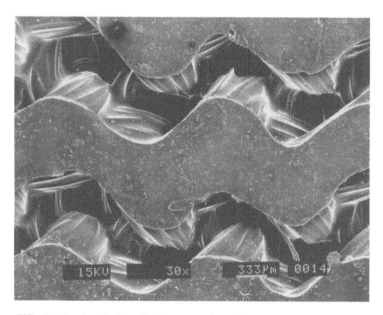

FIG. 7 The 1 × 1 rib weft knit composite, nylon multifilament, and epoxy resin; RTM processing.

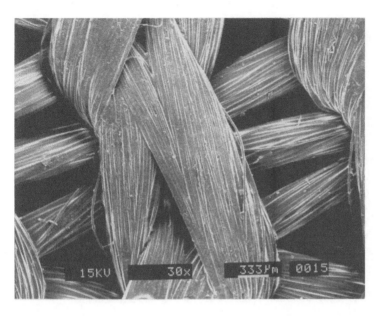

FIG. 8 The tricot warp knit composite, fiberglass multifilament, and epoxy resin; RTM processing.

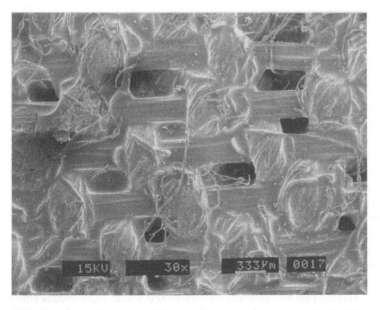

FIG. 9 The double cloth 102 composite, cotton warp, and Kevlar filling; RTM processing.

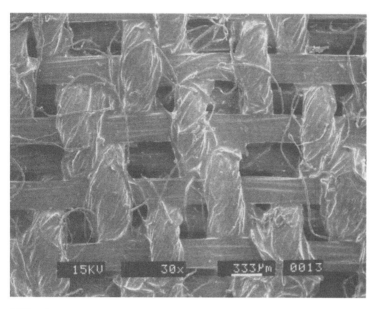

FIG. 10 The double cloth 201 composite, cotton warp, and Kevlar filling; RTM processing.

plastic matrix. The RTM parts showed that capillary flow will occur, as should happen with the thermoplastic matrix, provided that the viscosity becomes low enough for the molten polymer to infiltrate the yarn system.

In order to locally form the thermoplastic renitent composites, it is necessary to start with a yarn which contains both the load-bearing element and the thermoplastic element. Figure 11 shows some of the possible methods of commingling yarns or tows. Bicomponent filaments provide uniform mixing of thermoplastic within a yarn, but production costs are high. Bicomponent filament formation systems require simultaneous extrusion of two polymers; because a requirement of the renitent composite formation would be for one polymer to have a distinctly lower melting temperature than the other, this could add processing difficulties in addition to the added costs of coextrusion. Powder coating provides a good matrix dispersion, provided that the yarns are not overhandled, causing the powder to settle out of the system (a common problem during fabric formation using powder-coated yarns).

Commingled filaments give the yarn a more intimate mixing among the fibrous constituents. Having two sets of fibers equally mixed within yarns means either the filaments must be coextruded or each filament system be formed separately

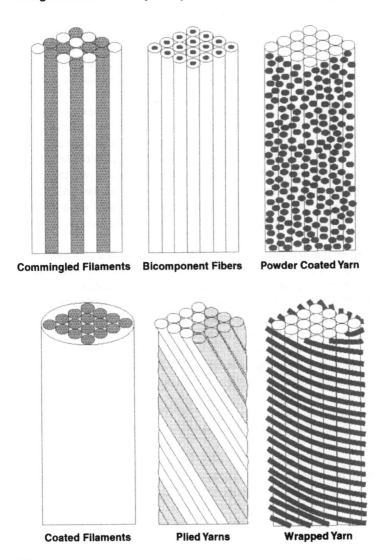

Commingled Filaments   **Bicomponent Fibers**   **Powder Coated Yarn**

Coated Filaments   **Plied Yarns**   **Wrapped Yarn**

FIG. 11 Forms of commingled yarns.

and later reeled together. The coating of yarns can be done by the dispersion of a coating matrix in a filament yarn (sometimes called "island in sea"). This process has become more economical in recent years and is less expensive than powder-coated or bicomponent filaments (provided the coating matrix is not of exorbitant cost).

Both plying and wrapping present economical solutions to commingling. Plying is done by providing a nominal twist to combine two (or more) sets of yarns. The wrapping process is realized by spinning staple fibers around a filament yarn core. The DREF-3 system is a friction spinning process which provides an efficient means of wrapping staple fiber around a continuous filament yarn to produce a commingled yarn [26]. Plying and wrapping do not provide the same degree of intermingling as do the other commingled alternatives, but if the end use is not stiffness intensive, this should not be a barrier to utilizing these economical systems.

All the yarns used in this portion of the study were composed of a continuous filament 195 denier Kevlar 49 yarn and a thermoplastic polyester component. The combination yarn was produced in two ways: with the DREF system and through plying. Plain jersey knit fabrics were made from these yarns. The properties of the yarns and subsequent fabric constructions are summarized in Table 2.

A micrograph of a section of the knitted fabric formed from DREF yarns is shown in Fig. 12. All that can be seen on the surface of these yarns is the polyester staple fibers, concealing the wrapped Kevlar core. The yarns appear very bulky upon observation; one constraint that may be encountered with the DREF spinning system is a lower limit in the amount of wrapper fibers in the formed yarn. At lower wrapper levels, the yarn tends to exhibit high variability, with many thick and thin spots. This problem is usually overcome at increased wrapper levels.

The fabric formed with plied yarns is shown micrographically in Fig. 13. The twisted nature of the plied yarns can be seen clearly. One yarn of the ply is Kevlar 49; the other yarn is a thermoplastic staple polyethylene terephthalate (PET) yarn. A problem visible in the photomicrograph is that the plies create preferential

TABLE 2 Weft Knitted Fabric Properties, Commingled Kevlar, and Polyester Yarns

Fabric	Combined yarn size	Yarn % Kevlar (by weight)	Courses per inch	Wales per inch	Fabric weight (g/m^2)
DREF	1000 denier	20%	24.3	16.7	390
Ply	550 denier	35%	21.5	16.3	200

FIG. 12 Jersey knitted fabric formed from DREF yarns, 80% Kevlar core with 20% polyester wrapper.

reinforcement areas, at the points where the thermoplastic ply twists around the reinforcement ply. The DREF yarns completely surrounded the Kevlar core without exhibiting the preferential reinforcement as in the plied yarns. The plied yarns will rely on capillary flow at low viscosity to achieve full encapsulation.

Composites were formed by applying heat and pressure to the knitted fabrics formed with commingled yarns. Increasing heating times were used to examine processing effects on the subsequent composite structure and porosity. The processing conditions used in this study are presented in Table 3, along with the physical properties of each produced composite system.

Figure 14 examines the relationship between the volume fraction (V_f) and processing time. It can be seen that as processing time lengthened, the volume fraction saw significant increase. It can be presumed that during the extra processing time, more of the thermoplastic matrix was allowed to infiltrate the fabric structure. A delicate balance must be reached, however: if the matrix is allowed to flow for too long a time, the interstices themselves may become enclosed. This can be seen for composite D3 (Fig. 17).

Photomicrographs for the produced renitent composites are contained in Figs. 15–20. For both plied and DREF yarns, the illustrations indicate that increased formation time results in lower porosity due to greater resin flow throughout the

FIG. 13 Jersey knitted fabric formed by plying 35% Kevlar and 65% polyester multifilament yarns.

TABLE 3 Consolidation Parameters and Physical Properties, Commingled Yarns

Sample	Consolidation time (s)/ heat (°C)/pressure (lbs)	Overall ρ (g/cm³)	V_f (% Kevlar)	Areal porosity (%)
DREF, dry	No consolidation	0.070	6.5	22.1
D1	15/275/5000	0.844	11.4	11.5
D2	30/275/5000	1.182	16.0	10.3
D3	45/275/5000	1.560	21.1	1.2
Ply, dry	No consolidation	0.025	8.9	34.4
P1	10/275/5000	0.477	11.8	25.2
P2	15/275/5000	0.546	13.5	20.0
P3	30/275/5000	0.926	22.8	4.9

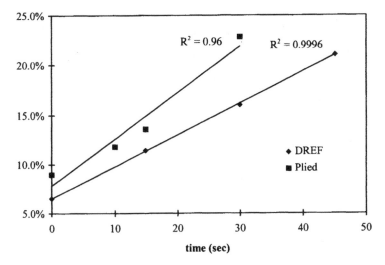

FIG. 14 Variation of volume fraction with processing time for composites formed from commingled yarns.

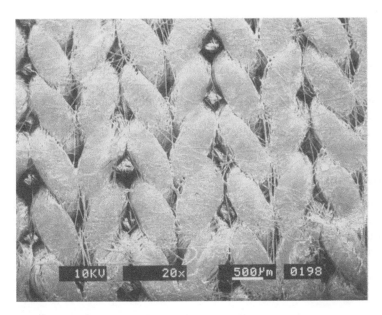

FIG. 15 Composite D1 formed from jersey knitted fabric of DREF yarns, 80% Kevlar core with 20% polyester wrapper.

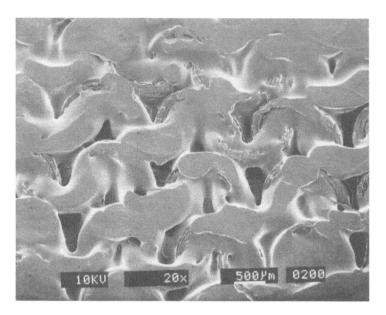

FIG. 16 Composite D2 formed from jersey knitted fabric of DREF yarns, 80% Kevlar core with 20% polyester wrapper.

composite. The plied yarns showed resin distribution mainly on the fabric surface, instead of flow into and around the Kevlar filaments. The DREF yarns showed better thermoplastic portion dispersion, exhibiting greater encapsulation of the Kevlar filaments than the plied yarns. Higher degrees of twist applied to the plied yarns should increase the encapsulation effect, but the structure of the DREF yarn is more directly tailored for achieving this effect. Sample P2 was not evaluated because it was nearly identical to P1.

D. Nonwovens

Flat nonwoven textile fabrics are here assumed to be a parallelepiped composed of air and fiber. The measurement of porosity is carried out by measuring the volume of the fabric and weighing the amount of fiber present. Using the density (ρ) of the fibers, the porosity can then be found from

$$\Phi = 1 - \frac{m}{V\rho} \tag{14}$$

where m is the mass of fibers and V is the volume of the parallelepiped fabric.

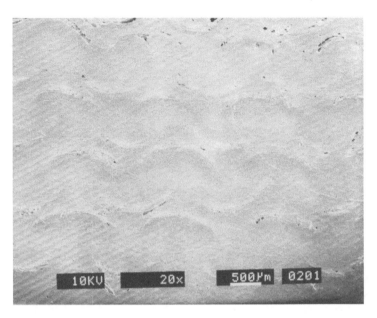

FIG. 17 Composite D3 formed from jersey knitted fabric of DREF yarns, 80% Kevlar core with 20% polyester wrapper.

In reality, a nonwoven fabric is not, of course, a perfect parallelepiped. The random arrangement of the fibers makes for surfaces exhibiting multiple ridges and valleys; even determining a minimum or maximum thickness value may prove difficult. Providing thermal surface bonding greatly improves the surface smoothness; this action was carried out for all of the fabrics produced for this project.

Nonwoven webs were produced using wet-laid and carded web techniques. The wet-laid webs consisted of a dispersion of polyester and Pulpex® fibers in a water suspension, mixed thoroughly. The water was then drained rapidly and flat nonwoven sheets were obtained on a drainage grid. Carded webs were made by carding a fibrous mat into a generally parallel web. Layers of the web were stacked with similar orientations to obtain parallel-laid webs and with orthogonal orientations to obtain cross-laid webs.

The webs produced are summarized in Table 4. The wet-laid webs were composed of 50% Pulpex and 50% thermoplastic polyester. The parallel- and cross-laid webs were 100% thermoplastic polyester. The polyester used in all webs was 1.5 denier, micrographically found to have a diameter of 12.4 μm. The Pulpex had varying linear density.

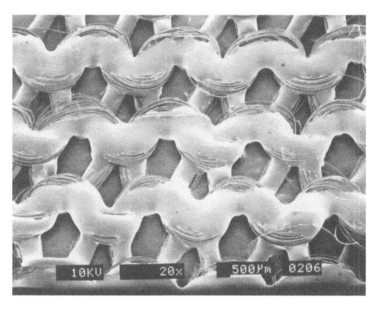

FIG. 18 Composite P1 formed from jersey knitted fabric of 35% Kevlar and 65% polyester plied multifilament yarns; face view.

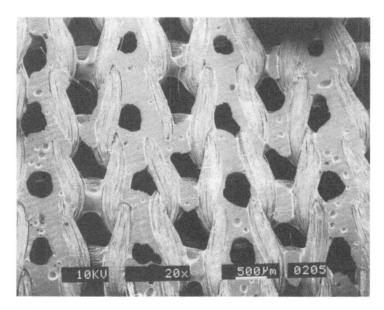

FIG. 19 Composite P1 formed from jersey knitted fabric of 35% Kevlar and 65% polyester plied multifilament yarns; back view.

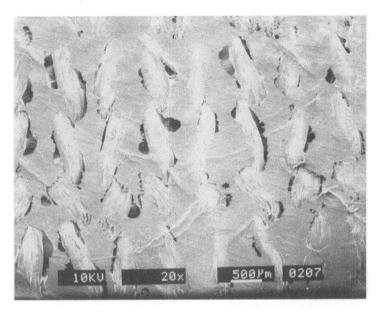

FIG. 20 Composite P3 formed from jersey knitted fabric of 35% Kevlar and 65% polyester plied multifilament yarns; back view.

All webs were produced both with and without spacers during thermal bonding in a heat press. The spacers used were thin metal tabs placed between the press plates whose presence limited the final part thickness to about 0.150 in. All webs were pressed at 200°F for 15 s. After consolidation, the web thicknesses were measured. The webs without spacers had much lower thicknesses than those with the spacers, with a minimum of half the thickness of parts formed without spacers. Using Eq. (14), the porosity for each web was found. These data are summarized in Table 5.

The webs produced with spacers had porosities that were significantly higher than those produced without spacers, directly attributable to the higher part thickness. The lower thicknesses resulted in the same amount of fiber having to be

TABLE 4 Nonwoven Web Characteristics

Web type	Wet-laid		Parallel-laid			Cross-laid		
Weight (g/m²)	180	340	50	120	280	50	150	220
Fibers	Pulpex & polyester		Polyester			Polyester		

TABLE 5 Nonwoven Fabric Properties

Web[a]	Thickness (mm)	Porosity (%)
w180n	0.203	22.1
w180s	0.889	82.0
w340n	0.559	47.8
w340s	1.22	75.4
p50n	0.152	75.3
p50s	0.635	94.1
p120n	0.292	66.7
p120s	1.52	94.3
p280n	0.584	64.5
p280s	1.93	88.7
x50n	0.178	78.1
x50s	0.457	90.2
x150n	0.381	70.7
x150s	1.43	90.3
x220n	0.432	63.2
x220s	0.838	80.6

[a] Code: aXXXb: a = w (wet-laid), p (parallel-laid), or x (cross-laid); XXX = web areal density (in g/m^2); b = n (no spacer used) or s (spacer used).

FIG. 21 Wet-laid nonwoven web of polyester and Pulpex, before thermal bonding.

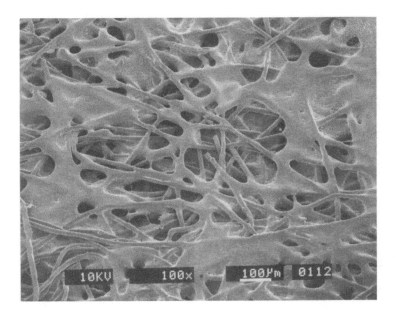

FIG. 22 Wet-laid nonwoven web of polyester and Pulpex, after thermal bonding.

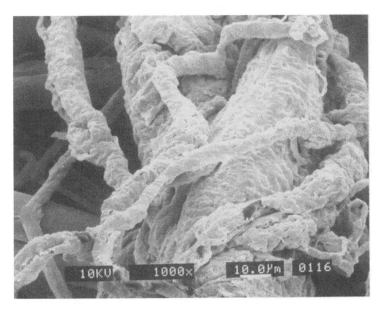

FIG. 23 Surface texture of Pulpex fibers.

FIG. 24 Parallel-laid nonwoven web of polyester fibers, before thermal bonding.

FIG. 25 Cross-laid nonwoven web of polyester fibers, before thermal bonding.

put into decreased volumetric space, yielding less space for air. This should yield higher permeability values (and lower resistance values), because an increase in the solid matter inside a parallelepiped should, in turn, have a greater resistance to the flow of air.

The porosities of the different web types can be examined in Figs. 21–25. The wet-laid web pictures are for the 340-g/m² weight, the parallel-laid for the 280-g/m² weight and the cross-laid for the 220-g/m² weight. All photomicrographs are for webs without spacers. By viewing these, it is evident that the wet-laid web had very little air space; even the surface regions that appear open are blocked by a second (back) layer of bonded Pulpex. The Pulpex had a very irregular cross section and surface (Fig. 23).

IV. TESTING

A. Kawabata Air Permeability Tester

Air permeability is measured as the resistance to airflow at a controlled velocity and pressure. Data were obtained using the Kawabata KES-F8-API Air Permeability Tester (Fig. 26). Air is pumped by a piston at a constant volume of 8π cm³/s. The velocity of the air is dependent on the plate chosen on the tester; each plate has a different aperture size, and air velocities of 0.4, 4, or 40 cm/s are possible. The pressure drop caused by the resistance of the specimen is measured by a differential pressure gauge. The output is the air resistance R, measured in kiloPascals times seconds per meter (kPa · s/m), found from

$$R = \frac{P_1 - P_2}{V} = \frac{\Delta P}{V} \tag{15}$$

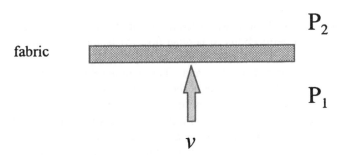

FIG. 26 Air permeability testing.

There are some defining equations of fluid flow that should be explained. If a specimen has small holes, the pressure drop is due to frictional loss and is defined as

$$\Delta P = KV \tag{16}$$

where ΔP is the pressure difference, V is the air velocity, and K is the constant for the specimen. This can also be expressed as

$$K = \frac{\Delta P}{V}$$

Here,

$$R = K$$

where R is resistance and is linear with respect to velocity for the specimen. A material with this response can be considered as a "linear resistor."

If the specimen exhibits large holes, then Bernoulli's law holds true, where

$$\Delta P = KV^2 \tag{17}$$

Rewritten,

$$KV = \frac{\Delta P}{V}$$

and

$$R = KV = \frac{\Delta P}{V}$$

Then, R is not constant because it is now a function of changing velocity. Such a material is considered as a "nonlinear resistor."

B.   Test Results

The composites formed by RTM were evaluated to identify the proper analytical technique to describe their permeability. If laminar flow exists, then the value of R should be constant for any V; if turbulent flow, R will vary with V.

The RTM samples were tested at 0.4, 4, and 40 cm/s in the Kawabata tester. Acoustic bomb testing was also carried out at the Northrop Grumman Corporation. This testing involves firing a frequency of sound at the media, with each frequency having some associated velocity. Similar to the Kawabata tester, the pressure drop across the media is measured and a resistance is reported based on Eq. (15). This testing provides a much greater range of velocities than the Kawabata tester, but it is considerably more expensive to perform. The data were

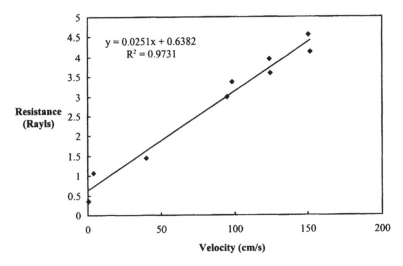

FIG. 27 Resistance versus velocity for tricot warp knit composite, fiberglass multifilament, and epoxy resin; RTM processing.

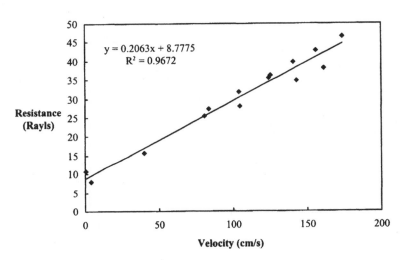

FIG. 28 Resistance versus velocity for double cloth 201 composite, cotton warp, and
Kevlar filling; RTM processing.

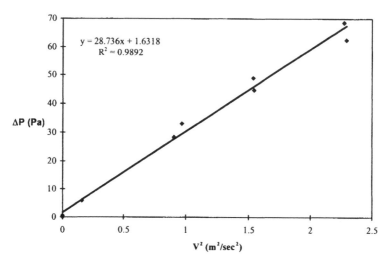

FIG. 29 Bernoulli fit for tricot warp knit composite, fiberglass multifilament, and epoxy resin; RTM processing.

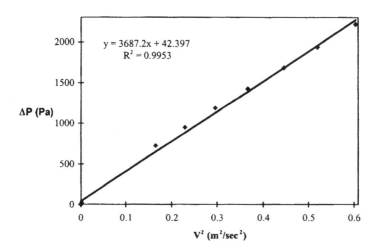

FIG. 30 Bernoulli fit for double cloth 201 composite, cotton warp, and Kevlar filling; RTM processing.

included because it includes high-velocity values, adding many points to each evaluated sample beyond the three provided by the Kawabata tester.

The traditional data report for permeability is to plot R versus V. This may be misleading if, in fact, R is inversely proportional to V^2. Figures 27 and 28 contain the R versus V plot with line fits for two RTM composites. If flow is laminar, then the line should have a slope of 0, an unvarying R.

Neither of the plots indicate laminar flow. In order to investigate the validity of turbulent flow (Bernoulli), the same data were plotted as ΔP against V^2 (Figs. 29 and 30). As can be interpreted from the better agreement (higher R^2 values), the data fit well to the Bernoulli equation. This technique may aid in understanding and modeling the permeability of porous media.

V. MODELING

A. Porosity Modeling

Permeability predictions using the findings of Darcy and MacGregor were carried out in this project. First, the data obtained from the RTM samples were evaluated.

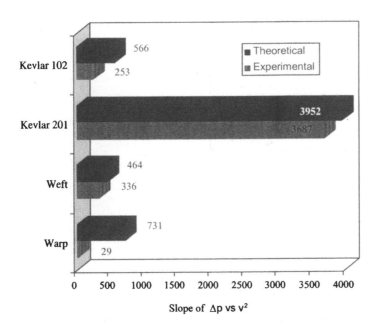

FIG. 31 Comparison of resistance slope and the MacGregor coefficient for RTM formed composites.

The slope of ΔP versus V^2 is R if there is turbulent flow. These slope numbers were compared to the predicted resistance as obtained by using the MacGregor equation. Values of B were predicted with measured values for the yarn diameter and composite porosity. Darcy assumed the resistance R of the filtration media to be laminar; by substitution, the Darcy equation becomes

$$B = \frac{\mu t}{R} \tag{18}$$

Using a value of 0.0175 cP (1.75×10^{-7} kPa $\cdot$ s) for the viscosity of air, the MacGregor prediction can then be used to predict the resistance of the media. This R is termed the MacGregor coefficient. Figure 31 compares the MacGregor coefficient to the slopes measured as in Figs. 29 and 30.

The MacGregor coefficient allows for a very general comparison of media resistances, but the prediction numbers appear to only allow accurate comparisons of media with distinctly different porosities. Varying velocity makes it difficult to describe media that obey turbulent flow.

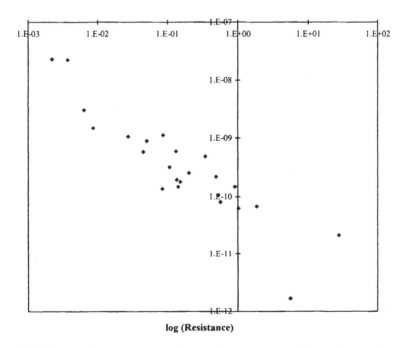

log (Resistance)

FIG. 32   Log(resistance) versus log(MacGregor) permeability prediction; all composite data.

B. Comparison with Results

Previous efforts have been centered on the evaluation of thick media with tortuous void structures, and image analysis was performed through the thickness direction to obtain a porosity value. Using the assumption that volumetric porosity values will be the same as areal porosity for thin sheets of uniform cylindrical cavities, image analysis of the surface geometry will suffice for renitent composites. Hence, image analysis was used to obtain areal porosity values as acquired in the respective sample photomicrographs.

A global comparison involving all formed renitent composites may be achievable by comparing the measured resistance of all samples at a constant velocity. This was done by evaluating each renitent composite at a test velocity of 4 cm/ s and plotting the resistance measurement against the predicted permeability from the MacGregor equation on a log–log plot (Fig. 32). There is a general linear fit for the data when evaluated with this technique; however, it is difficult to evaluate negative numbers on log plots, yielding the need for a refined method.

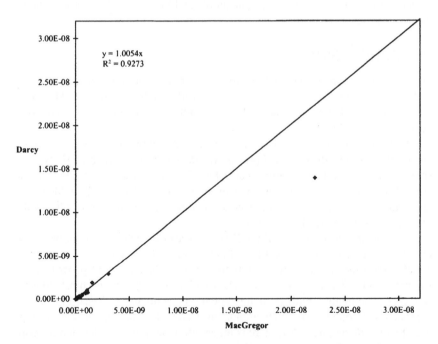

FIG. 33 Comparison of MacGregor and Darcy permeability predictions for all composite data.

Each renitent composite was also evaluated for the predicted permeability using Eq. (15). Figure 33 compares the predicted permeabilities found using both Darcy and MacGregor. The line represents perfect agreement between the two predictions. There is a very good association at lower permeability values; the overall agreement is still very good, with an R^2 value of 0.9273. The two outliers illustrate that extreme values of permeability still show variation. This could be due to the existence of turbulent flow, at which the MacGregor and Darcy interpretations seem to separate.

C. Discussion of Results

The results as observed in Figure 33 are encouraging; they imply that for a number of different structures, their permeability trend can be predicted with some degree of confidence. This comparative technique could prove quite useful in application, as many times a prototype is developed with an established product identified as a performance target. Simple evaluation of different variables that may be involved in the prototype production can be input into the Darcy and MacGregor relationships, and the effect of each parameter can be quickly observed. This scheme of ''ranking'' materials and their variables is a reasonable means of product evaluation.

VI. CONCLUSIONS

Porous barriers can be successfully manufactured by the formation of renitent textile composites. Composites were successfully produced using RTM and by thermoplastic matrix incorporation. Woven, knitted, and nonwoven renitent composites were produced.

The DREF yarn spinning system seems to hold promise, both for its economy and utility. Further adjustment of processing and composite formation techniques should lead to excellent filtration media based on these yarns. The need for the cross-discipline technology transfer between textiles and materials industries is apparent. There is a need for improved processing to ensure uniformity in the production of wide width materials.

The MacGregor derivation of the Kozeny–Carman equation was shown to be on the same magnitude as direct use of Darcy's law. Differences in the numbers could come from many sources, including inexact thickness and porosity measurement and differing air viscosity. The trend as identified by each relationship was very encouraging, with good agreement on a comparative basis. The simple approach of a comparative evaluation using Darcy and MacGregor relationships is fast and valuable; however, there is much left unanswered. A detailed description of turbulent flow is necessary if renitent composites with micropores can be made. This would require in-depth evaluation of the unique pore geometries formed by the contained fabrication techniques.

REFERENCES

1. F. Ko, C. Pastore, and A. Head, *Atkins and Pearce Handbook of Industrial Braiding*, 1989, pp. 2.21–2.23.
2. D. Williams. *Ad. Compos. Eng.* 12 (June 1987).
3. F. Ko, K. Krauland, and F. Scardino, in *Progress in Science and Engineering of Composites*, 1982, pp. 1169–1176.
4. A. Bogdanovich and C. Pastore, *Mechanics of Textile and Laminated Composites*, Chapman & Hall, London, 1996.
5. B. Cox. J. Compos. Mater. 1114 (1994).
6. GCA Corporation, *Handbook of Fabric Filter Technology*, GCA Corp., 1970.
7. B. Miller, et al., *Influence of Fiber Characteristics on Particulate Filtration*, GCA Corp., 1975.
8. C. Dickenson, *Filters and Filtration Handbook*, Elsevier Science, London, 1992.
9. H. Darcy, *Les Fontaines Publiques de la Ville de Dijon*, Paris, 1956.
10. J. Daily and D. Harleman, *Fluid Dynamics*, Addison-Wesley, Reading, MA, 1966, pp. 180–184.
11. A. Scheidegger, *The Physics of Flow Through Porous Media*, rev. ed., University of Toronto Press, Toronto, 1960.
12. J. Berryman and S. Blair. J. Appl. Phys. 2221 (1987).
13. R. McGregor. J. Soc. Dyers Colour. *81*:429 (1965).
14. F. T. Pierce. Textile Res. J. *17*(3):123.
15. S. Ergun. Chem. Eng. Prog. *48*(2):89 (1952).
16. M. Cheikhrouhou and D. Sigli. *Textile Res. J.* 371 (1988).
17. T. Starr. J. Mater. Res. 2360 (1995).
18. J. Wagner, *Nonwoven Fabrics*, Philadelphia College of Textiles and Science, Philadelphia, 1982, pp. 46–47.
19. L. Van Den Brekel and A. De Jong. Textile Res. J. 433 (1989).
20. A. Mahale, R. Prud'homme, and L. Rebenfeld, in *Fiber-Tex 1992: The Sixth Conference on Advanced Engineering Fibers and Textile Structures for Composites*, 1993, pp. 183–202.
21. G. Griffiths, et al. SAMPE J. *20*(32) (1984).
22. *Engineered Materials Handbook, Vol. 1: Composites*, ASM International, Metals Park, OH, 1987, pp. 97–104.
23. S. Hasselbrack, C. Pederson, and J. Seferis. *Polym. Compos. 13*(1) (1992).
24. J. Mayer, et al. *Automot. Eng.* 21 (August 1994).
25. T. Kawase, et al., *Textile Res. J.* 409–413 (1986).
26. P. Lord, *Economics, Science & Technology of Yarn Production*, School of Textiles, North Carolina State University, 1979.

Index

T - #0038 - 111024 - C0 - 229/152/18 - PB - 9780367397869 - Gloss Lamination